Industrial Biofouling

Industrial Biofouling

DETECTION, PREVENTION AND CONTROL

Edited by

JAMES WALKER
CAMR, Porton Down, Salisbury, Wiltshire, UK

SUSANNE SURMAN
Central Public Health Laboratory, London, UK

JANA JASS
Department of Microbiology, Umeå University, Sweden

JOHN WILEY & SONS, LTD
Chichester • New York • Weinheim • Brisbane • Singapore • Toronto

Other Wiley Editorial Offices

John Wiley & Sons, Inc., 605 Third Avenue,
New York, NY 10158-0012, USA

WILEY-VCH Verlag GmbH, Pappelallee 3,
D-69469 Weinheim, Germany

Jacaranda Wiley, Ltd., 33 Park Road, Milton,
Queensland 4064, Australia

John Wiley & Sons (Asia) Pte, Ltd., 2 Clementi Loop #02-01,
Jin Xing Distripark, Singapore 129809

John Wiley & Sons (Canada) Ltd., 22 Worcester Road,
Rexdale, Ontario M9W 1L1, Canada

Cover
Epifluorescent microscopy photographs of 5-d-old Pseudomonas fragi biofilms, which were stained
with acridine orange. The biofilms had been grown on various stainless-steel surfaces (AISI 304) glass
blasted, lapped and mechanically polished surfaces. Reproduced by the kind permission of Gun
Wirtanen, VTT Biotechnology and Food Research, Espoo, Finland.

Library of Congress Cataloging-in-Publication Data
Industrial biofouling : detection, prevention, and control / edited by James Walker,
Susanne Surman, Jana Jass.
    p.cm.
    Includes bibliographical references and index.
    ISBN 0-471-98866-9 (alk. paper)
    1. Fouling. 2. Fouling–Measurement. 3. Fouling organisms–Control. 4. Industrial
water supply. I. Walker, James (James T.) II. Surman, Susanne. III. Jass, Jana.
TD 427.F68 I58 2000
628.1′683–dc21
                                                                            00-038174

British Library Cataloguing in Publication Data
A catalogue record for this book is available from the British Library

ISBN 0 471 98866 9

Typeset in 10/12pt Palatino from the authors' disks by Dobbie Typesetting Limited, Devon
Printed and bound in Great Britain by Biddles, Guildford, Surrey
This book is printed on acid-free paper responsibly manufactured from sustainable forestry,
in which at least two trees are planted for each one used for paper production.

Dedication

We would like to dedicate this book to 'our supportive families'
including Jordan & Per-Erik, Gill, Philip & Richard and
John, Nicola and David as well as to all those that inspired us
to study biofilms.

Contents

Contributors

Dr Anne Camper, Center for Biofilm Engineering, Montana State University, Bozeman, Montana, 59717, USA.

Professor Hans-Curt Flemming, Gerhard-Mercator-Universität-GH, Duisburg Moritzstraße 26, 45476 Mulheim an der Ruhr, Germany.

Professor Joe Frank, University of Georgia, Department of Food Science, College of Agriculture Athens, GA 30602, USA.

Dr Liz Fricker, Thames Water Utilities, Spencer House Laboratory, Manor Farm Road, Reading, RG7 1HD, UK.

Dr Thomas Griebe, Gerhard-Mercator-Universität-GH, Duisburg Moritzstraße 26, 45476 Mulheim an der Ruhr, Germany.

Dr Jana Jass, Umeå University, Department of Microbiology, Building 6L, SE-901 87 Umeå, Sweden.

Dr Martin Jones, Unilever, Portsunlight Research, Quarry Road, East Bebbington, Wirral, L63 3JW, UK.

Dr Tiina Mattila-Sandholm, Research Scientist, VTT Biotechnology, Tietotie 2, Espoo, PO Box 15012028, FIN-02044, Finland.

Dr Gordon McFeters, Department of Microbiology, Montana State University, Bozeman, Montana, 59717, USA.

Professor Glyn Morton, Department of Biological Sciences, University of Central Lancashire, Preston, UK.

Dr Steven Percival, Department of Biological Sciences, The University of Central Lancashire, Preston, PR1 2HE, UK.

Dr Erna Storgård, Research Scientist, VTT Biotechnology, Tietotie 2, Espoo, PO Box 15012028, FIN-02044, Finland.

Dr Maria Saarela, Research Scientist, VTT Biotechnology, Tietotie 2, Espoo, PO Box 15012028, FIN-02044, Finland.

Dr Satu Salo, Research Scientist, VTT Biotechnology, Tietotie 2, Espoo, PO Box 15012028, FIN-02044, Finland.

Dr Susanne Surman, London Food, Water and Environmental Microbiology Unit, Central Public Health Laboratory, London NW9 5HT, UK.

Dr Joanne Verran, Department of Biological Sciences, Manchester Metropolitan University, Chester Street, Manchester, MMM 5GD, UK.

Dr James T Walker, CAMR, Porton Down, Salisbury, Wiltshire, SP4 0JG, UK.

Dr Gun Wirtanen, Senior Research Scientist, VTT Biotechnology, Tietotie 2, Espoo, PO Box 15012028, FIN-02044, Finland.

Preface

A broad spectrum of disciplines have been covered in this book, including drinking water systems, industrial water systems and food industries, in which biofilms are a persistent problem. In the water service utilities, biofilms not only affect the taste and potability of the water but may also result in cross infection problems through regrowth of coliforms. Industrial water systems such as cooling towers and engineering processes, including paper mills, have their own problems due to the manifestation of biofilms that clog filters which lead to increased costs due to decreased flow and increased pipeline maintenance and can result in occupational exposure. In engineering, biofilms cause problems such as degradation of cutting fluids due to biofilm metabolism and growth, corrosion and increased frictional resistance that result in increased costs and also an increased risk of workplace acquired infections. Biofilms often result in the end spoilage of a food product and this is a particular problem in the food preparation and manufacturing industries.

This book provides an overview of current research with the chapters containing three sections; (i) problems associated with biofilms; (ii) how to detect them, and (iii) control mechanisms. Individual experts have written each section with the authors ranging from all over the world including Europe, USA and Canada. The information contained will provide insight into each of the areas for lay persons, students and graduates whilst giving current research from which the specialist can obtain the most relevant up to date research information available.

A companion volume describing biofilms in medicine is currently in preparation. This book will discuss the problems, detection and control of biofilms associated with medical devices and implants, tissue surfaces and dental materials.

Glossary

Biofilm The formation as a result of microorganisms attached to an interface and growing within a glycocalyx matrix.

Biofouling (Biological fouling) The accumulation and growth of living organisms and their associated organic and inorganic material on a surface.

Biodeterioration The deterioration of materials of economic importance by microorganisms. Undesirable change or degradation in the properties of a material caused by the metabolic activities of microorganisms.

Biologically influenced corrosion (BIC) The corrosion of a metal as a result of biological activity and substances.

Biotransfer potential Term used in the food industry when determining the risk of contamination of the product or surfaces by microorganisms from another contaminated surface.

Chemical reaction fouling The fouling deposits on a surface as a result of chemical reactions within the aqueous phase that do not involve the surface as a reactant.

Coliforms Group of bacteria that are used as indicator organisms for the presence of faecal contamination and the presence of pathogens.

Conditioning film The film of material adsorbed to a surface when it is placed into an aqueous environment. The film often alters the characteristics of the surface and thus influences microbial adhesion.

Corrosion The chemical reversion of a refined metal to its most stable energy state often causing surface deterioration and pitting.

Covalent bonds A bond created between atoms by sharing electrons.

Disinfection The killing of microorganisms, often by chemical means such as antibiotics or biocides.

Electrostatic interactions The forces incurred between charged groups. The attractive forces between opposite charges and the repulsion forces between like charges.

Exopolysaccharides (EPS) Polysaccharide containing material or slime substances, secreted by some organisms when in association with a surface and often involved in biofilm formation. Slime substances may be composed of polysaccharides, proteins, nucleic acids and lipids.

Extracellular polysaccharide substances (EPS) See Exopolysaccharides.

Fouling   The undesirable formations of organic and/or inorganic deposits on surfaces.

Glycocalyx   An exopolysaccharide-containing material of microbial origin, outside of the bacterial membrane or cell wall. Glycocalyx is identified as the main component surrounding microbial communities within a biofilm.

Hydrogen bonds   The bonds formed due to the attraction of a hydrogen atom (i.e. from a hydroxyl group) to an electronegative atom on another molecule (i.e. oxygen).

Heterotrophs   Microorganisms that require complex organic molecules as carbon nutrient and energy source.

Microbial corrosion (MC)   Corrosion of a metal surface caused by microbial activity and their metabolic by-products.

Microbial influenced corrosion (MIC)   See microbial corrosion.

Nosocomial infections   Hospitally acquired infections. These are often resistant to many antibiotics.

Opportunistic pathogens   Microorganisms that are not normally pathogenic and are part of the natural microflora, however, when displaced, they become pathogens and may cause serious infections.

Planktonic   Microorganisms that are freely dispersed within the aqueous phase, for example planktonic bacteria refers to bacteria suspended in a liquid medium.

Sessile   Microorganisms that are attached to an interface or surface and within a biofilm.

Sloughing   A term used in the study of biofilms for the loss of surface biofilm material due to hydrodynamic force.

Soiling   Fouling of food contact surfaces by organic material in the absence of microorganisms.

Quorum sensing   The sensing of a critical population density by microbial cells. Microorganisms monitor their population density by communication molecules.

Van der Waals Interactions   The weak attraction forces that occur between atoms or molecules. For example the attraction between hydrophobic molecules to each other.

Viable but non-culturable (VNC)   Bacteria that are viable under the optimal conditions, however, they do not grow under standard culture techniques. These organisms are regarded as being in a state of dormancy.

1 Biofilms and Biofouling

J. JASS[1] and J. T. WALKER[2]
[1]*Project leader, Department of Microbiology,*
Umeå University, SE-901 87 Umeå, Sweden
[2]*Microbiologist, Pathogen Microbiology Group, CAMR,*
Porton Down, Salisbury SP4 0JG, UK

INTRODUCTION

In the environment, microorganisms grow in two distinct forms; as sessile cells within a biofilm attached to an interface or surface and as planktonic organisms freely dispersed in the aqueous phase (Marshall 1994). Biofilms are adherent microbial communities surrounded by a matrix composed of microbial polymer and components originating from the environment (Characklis and Marshall 1990; Lappin-Scott *et al.* 1992). A wide range of microorganisms, including bacteria, protozoa, algae and fungi, have been associated with biofilms. These adherent colonies are often located on surfaces within aqueous environments, but may also be found at air–liquid or liquid–liquid (phase separation) interfaces (Bar-Or 1990). Biofouling is the accumulation of organic material and debris on a surface and often includes the presence of microorganisms. In most environments microorganisms prefer to grow as sessile communities and this is likely due to the protective nature of biofilm growth (Geesey *et al.* 1978).

Microbial growth, in the form of biofilms, has been found to be ubiquitous in nature, medicine and many industries (Jass and Lappin-Scott 1997). Adherent microbes are involved with the biofouling and degradation of surfaces in many industrial processes including food, drink, water service utilities (Carpentier and Cerf 1993; Block *et al.* 1993), pharmaceutical, paint, oil (Bar-Or 1990), paper (Väisänen *et al.* 1994), manufacturing and engineering industries (Challinor 1991), in addition to the health related fields of medicine and dentistry. The presence of microbial biofilms in such a wide range of habitats indicates a need for an understanding of their presence and their potentially harmful effects within individual environments.

Industrial Biofouling. Edited by J. Walker, S. Surman, J. Jass
©2000 John Wiley & Sons Ltd

BIOFILM FORMATION

Biofilm formation occurs in a number of stages: the formation of a conditioning film; microbial association with the surface; adhesion to the surface; growth and proliferation; recruitment of other microorganisms (Figure 1.1); and finally detachment (Brading *et al.* 1996). In different industrial environments, each stage of biofilm formation will have different characteristics, however, here we will describe biofilm formation on a solid surface in an aqueous environment such as on a water distribution pipe. Specific features of the biofilms found in different industries are described later in this book.

A clean surface placed into any aqueous environment will rapidly become coated with a layer of ions and organic molecules called a conditioning film or layer. This conditioning layer functions by altering the surface properties and providing a source of nutrients in oligotrophic environments (Marshall 1996). The microorganisms will then become associated with the surface, possibly being attracted to the accumulated ions and other potential nutrients. The cells will initially adsorb loosely to the surface in a reversible manner so they can become easily detached. At the surface, the microorganisms synthesize large amounts of exopolymer which leads to stronger irreversible attachment. Irreversible attachment may occur non-specifically, by van der Waals interactions, electrostatic forces and hydrogen bonds and/or by specific receptor mediated interactions and chemical bonds (Vandevivere and Kirchman 1993; Marshall 1992, 1994). When securely attached to the surface, the microbes can grow and divide to form microcolonies, which in time develop into mature biofilms possessing a three dimensional structure (Figure 1.2).

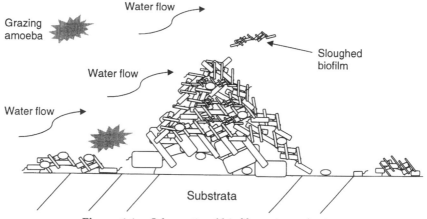

Figure 1.1. Schematic of biofilm microcolonies.

At a critical stage of biofilm maturity, microbial components may be released, as single organisms or as large particles, into the bulk aqueous phase (Stewart *et al.* 1993). This maintains the biofilm at an equilibrium or steady state dictated by the organisms, nutrient availability and environmental conditions. Some organisms produce substances that degrade the polymer matrix to aid in the release of cells from the biofilm in response to their environmental conditions (Lee *et al.* 1996; Boyd and Chakrabarty 1994). Stewart *et al.* (1993) found that pure culture *Pseudomonas aeruginosa* biofilms shed particles up to 100 μm^2 in size, suggesting that recontamination of industrial processes would be high if the biofilm was not completely removed. It is not always the formation of a biofilm that causes problems, but the shedding and detachment of biofilm particles that may produce the deleterious effects.

BIOFILM STRUCTURE

Biofilm structure is dictated by the environment in which the microorganisms are residing, the nutrients available, the organisms present, the physico–chemical properties of the interface and the hydrodynamics of the

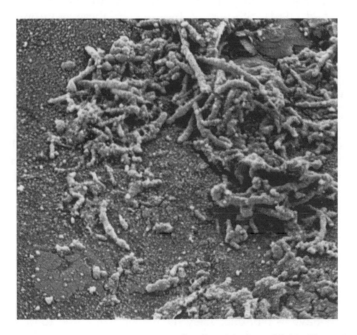

Figure 1.2. Scanning electron micrograph of a 14 day old biofilm microcolony growing out from the substrata (2200 × magnification).

surrounding bulk phase (Stoodley *et al*. 1999; Lawrence *et al*. 1991; Costerton *et al*. 1994; Korber *et al*. 1995). The wide range of environments that microbes encounter within different industries and processes, suggest that very different biofilm structures exist. However, the different microbial structures still adhere to the criteria of a biofilm; microorganisms associated with an interface and surrounded by a polymer matrix.

A major part of the biofilm structure is the exopolysaccharide matrix, produced by the cells but often containing both organic and inorganic components from the surrounding environment (Characklis and Marshall, 1990). Optical-sections of biofilms have shown that extracellular material and water compose approximately 98% of the biofilm structure and this is referred to as the glycocalyx (Lawrence *et al*. 1991; Costerton *et al*. 1992). It functions in both the early stages of biofilm formation, where the polysaccharide matrix is required for adhesion, and the later stages, where it provides structure for the microorganisms to reside. The glycocalyx is important, not only for containing the cells within a suitable environment, but also for protecting the cells from antimicrobial substances (Costerton *et al*. 1992, 1994).

In water distribution pipes (Block *et al*. 1993), heat exchangers and other conduits (Challinor 1991), a major influence on the biofilm structure is the low nutrient status and potentially high hydrodynamic shear forces (Stoodley *et al*. 1994). Studies of biofilms in these environments have provided a basic biofilm model to which others have been compared. This model biofilm is a highly hydrated and porous structure containing microbial cell clusters that are separated by voids and channels. The voids and channels contain liquid, at a much slower flow rate than in the bulk phase, while the cell clusters are held together by a polysaccharide matrix. Under high flow rates, biofilms also contain streamers which are long strands of glycocalyx and microbes dragged down stream in the aqueous phase (Stoodley *et al*. 1999). Keevil and Walker (1992) found that biofilms in cooling towers contained similar structures, where tall stacks of bacteria in polysaccharide extended far into the aqueous phase. However, these stacks were anchored to a basal cell layer covering the surface rather than directly to the surface. The stacks, cell clusters and streamers are believed to reach far into the bulk fluid providing a large surface area and volume in order to sequester any available nutrients in the bulk phase.

In high nutrient environments, where there is a copious amount of debris, such as in many food processing industries, biofilms may become thick and packed with both microorganism and debris (Mattila-Sandholm and Wirtanen 1992). While it may be suggested that these are not biofilms, the structures behave as biofilms. They have been compared to dental plaque, where there is generally high nutrient levels, with the exception of oxygen. Oxygen is likely to be the limiting nutrient, however, many of the microorganisms are facultative and can grow under anoxic or anaerobic conditions. These conditions will determine which microorganisms can survive within these thick and highly stratified biofilms. Often in the food processing industry, biofilms are not allowed to reach these levels, as they pose a threat, not only, to the quality of the product but also to the health of the consumer. The biofilms that develop in these areas are those in drains and on the walls of processing plants (Mattila-Sandholm and Wirtanen 1992).

While biofilm structure is important, the key feature to biofilms is that they are highly dynamic, multidimensional, structures that are continuously changing in time and space (Wimpenny et al. 1993). This dynamic nature of the biofilm is a response by the organisms to continuously changing environments providing a better chance of survival.

ADVANTAGES OF GROWTH WITHIN BIOFILMS

There are a number of advantages that result from microorganisms growing in communities at interfaces. That is, they have increased initial available nutrients, nutrient cycling and exchange, protection from antimicrobial substances, desiccation and scavenging, creating desirable micro-environments under generally harsh conditions, plasmid transfer and cell communication. While different environments will emphasize different advantages, they all play a role in the selection for biofilm forming organisms.

In oligotrophic environments, microbial exopolymer may be used for accumulating and sequestering ions and small charged molecules for use as nutrients by the microorganisms. The glycocalyx has previously been characterized as repeating units of polymer with a net negative ionic charge and may behave as an anionic exchange column, trapping cations and positively charged substances (Costerton et al. 1992). Additionally, other substances, such as plant material and cellulose, may be trapped physically and used as a nutrient source. While a nutrient to one microorganism may not be to another, nutrient exchange may be required for microbial survival. This leads to communities composed of different microorganisms with complementary needs. The degradation of substrates to simple molecules, by one organism, will be used as a nutrient source, by another. For example, the cellulose degrading *Clostridium thermocellum*, degrades cellulose into simple disaccharides that may be used as nutrients by other organisms residing in the biofilm (Bayer and Lamed 1996). This type of exchange is possible for cells living in close proximity to each other as found within biofilms. In the case of low nutrient environments, some organisms may starve and lyse providing others with nutrient by cryptic growth (Banks and Bryers 1990). Thick biofilms will have nutrient gradients, especially for oxygen, where the upper regions will be aerobic, the lower regions will become anoxic and finally anaerobic. Such biofilms would likely be found in the food industry if allowed to develop.

Sessile microorganisms are more resistant to antimicrobial and biocidal agents, than when they are growing as planktonic cells (Keevil *et al.* 1990; deBeer *et al.* 1994; Korber *et al.* 1994). Laboratory and *in situ* studies have shown the survival of a large number of organisms after chemical treatment (Jass and Lappin-Scott 1994; Walker *et al.* 1994). Protection of the microorganisms may occur in a number of different ways. The biofilm may prevent the biocide from reaching the inner-most cells by the presence of the exopolysaccharide matrix (Suci *et al.* 1994). The microorganisms within the biofilm may have a reduced growth rate, therefore, the slowly metabolizing organisms will be less affected by some agents (Brown *et al.* 1988; Stewart *et al.* 1994). The biofilm environment may produce an altered phenotypic response that reduces antimicrobial efficacy. Finally, some biofilm cells may have genetic resistance, i.e. on a plasmid, that may protect them and perhaps this may be transferred to other organisms within the biofilm.

Microorganisms growing within a biofilm are not only protected from antimicrobial agents, but also from harsh environmental conditions. These may be intermittent drying–wet cycles, high or low temperatures, UV irradiation, low nutrient conditions and high hydrodynamic shear rates. The highly hydrated microbial exopolysaccharide surrounding the micro-organisms protects them from desiccation (Ophir and Gutnick 1994; Roberson and Firestone 1992). The biofilm structure may, to some extent, protect the sessile organisms from predation, although, amoeba have been found feeding on biofilm covered surfaces (Walker *et al.* 1991).

The nature of the biofilm environment suggests that plasmid transfer between organisms would occur, since the donor and recipient organisms are in close association with each other. Although this has been difficult to prove directly *in situ*, a number of studies have shown that transconjugation does occur under certain conditions. Angles *et al.* (1993) found that using antibiotic resistance markers on plasmids, transconjugation between a *Vibrio* sp. and a marine bacterium occurred more often in biofilms than in the planktonic phase after a minimum of 170 h. Hausner and Wërtz (1999), using the green fluorescent protein reporter gene to detect transconjugants, found that the frequencies were dependent on biofilm structure and contact time. This is of some concern for industries where antibiotic resistant plasmids may be found, such as in the food and pharmaceutical industries.

Microorganisms produce small diffusable molecules that allow them to communicate with each other, called quorum sensing. These molecules have been identified as homoserine lactones for Gram-negative organisms and peptides and other small molecules for Gram-positive bacteria and other species (Fuqua *et al.* 1994). These substances are produced by microorganisms to regulate their population density by accumulating to a critical concentration at high densities, which then elicits a genetic response from the cells. For example, *P. aeruginosa* contains 2 genes, producing 2 different homoserine lactones, that are involved in quorum sensing. By creating mutants, Davies *et al.* (1998) found that these molecules do not affect bacterial attachment, however, subsequent biofilm formation is blocked. Furthermore, this system also regulates the expression of virulence factors such as pili, which have also been shown to be important in adhesion to abiotic surfaces (Pratt and Kolter 1998). *Escherichia coli* and *Salmonella typhimurium* have similar quorum sensing systems to *P. aeruginosa* (Surette and Bassler 1998) and environmental microbial isolates have been observed to produce these quorum sensing molecules *in situ* (McLean *et al.* 1997).

PROBLEMS OF BIOFILMS AND BIOFOULING IN INDUSTRY

Biofouling by microorganisms can disrupt the normal operation of many industrial systems, primarily where water is used in the process. This results in high costs to the industry due to the reduced process efficiency, clean-up and shut down time. It is also disconcerting that these biofilms may pose a major health risk (Challinor 1991). It is obvious that biofouling and microbial contamination can have health risks in the food and drinking water industries if contaminated food is consumed. However, it is also possible that the workers, exposed to microbial pathogens residing in biofilms during processing, would be at risk (Carpentier and Cerf 1993; Block *et al.* 1993). For example, in industrial cutting fluids where water/oil emulsions are used as lubricants, microbial contamination at the oil–liquid interface may not only degrade the cutting fluid but may also infect the respiratory tract or wounds of the workers.

In the food and drinking water industries, the microbial contamination may be involved in food spoilage by introducing undesirable tastes or smells to the end product (Carpentier and Cerf 1993). They also provide a site for pathogens to reside and grow, causing health risks to the consumer. In addition, the workers are exposed to these pathogens since the cleaning of food contaminated areas produce aerosolized particles containing microorganisms that may cause respiratory infections. Bacterial adhesion is believed to be a virulence indicator. Solano *et al*, (1998) tested different isolates of *Salmonella enteritidis* from the environment, dairy products and patients, for their ability to form biofilms and found that only the virulent strains were able to form thick biofilms. Microorganisms residing within a biofilm are resistant to high levels of antimicrobial/biocide and harsh environments, such as desiccation, thus, often surviving cleaning and disinfecting regimes. If surfaces are not sufficiently cleaned, re-growth and cross-contamination results.

In oligotrophic or low-nutrient environments, microorganisms seem to prefer to grow on surfaces and at interfaces. Potable water distribution systems are often perceived to be hostile environments, completely free of nutrients and organic carbon, low temperatures, high flow rates and residual amounts of disinfectant (Camper et al. 1999; Block et al. 1993). It is perceivable to think that nothing would grow under such conditions, but microorganisms have evolved so that they are masters at extracting and sequestering nutrients. Molecules and ions are adsorbed to surfaces, which makes them available for surface microorganism. They attach and produce excessive amounts of exopolymer, which assists in protecting the cells from antimicrobial substances and hydrodynamic sheer forces. Biofilms can trap other species of microorganisms such as coliforms and opportunistic pathogens such as *Legionella pneumophila* and *P. aeruginosa* (Camper et al. 1996).

In all of the industries mentioned, biofouling of surfaces by microorganisms may cause degradation and corrosion of the surfaces. In the food industry, it produces pitting and thus difficulties in subsequent cleaning of the surfaces resulting in faster re-growth times. In the water and oil industries, biofouling causes corrosion of pipes and reduced product quality. Microbial corrosion of metal surfaces under certain conditions may corrode through a 2 cm steel plate within 6 months (Lappin-Scott and Costerton 1989). The primary organisms that cause corrosion are anaerobic sulphate reducing bacteria. A corrosion cell may be created in two ways. First, the bacteria produce a cathode by releasing organic acids as fermentation products, thus a corrosion potential is created due to the differential in proton ions (Hamilton 1995). The second process that creates a corrosion cell is when metal ions are released by anionic polysaccharides, creating an unequal cation distribution across the surface (Bremer and Geesey 1991; Geesey et al. 1987; Walker et al. 1991).

Biofouling of industrial processes is often in the form of microbial contamination on surfaces and the release of microorganisms into the surrounding process fluid. Although microbial accumulations at interfaces, such as air–liquid and at liquid–liquid phase separations, may seem quite different, they are not when identifying the major features associated with biofilms. For example, an interface provides a site where both nutrients and microorganisms may accumulate. Biofilms that are formed at these interfaces are surrounded by exopolysaccharide material of microbial origin and may include substances trapped from the bulk fluid. This structure behaves in the same manner as a biofilm on a surface; by protecting the resident microbes from antimicrobial activity, by providing a means of nutrient entrapment and in the case of mixed communities, by allowing nutrient exchange. Although we focus on biofilms and fouling of solid surfaces, it is important to remember that biofilms formed at interfaces may pose the same health risks and similar problems in the process and food industry as those on solid surfaces.

CONTROL OF BIOFILM

There are two major strategies for biofilm control: prevention or cleaning and disinfecting (sanitizing). Biofilms have been found to eventually colonize most surfaces under hydrated conditions. Therefore, prevention may be difficult with the present technology, however, different surfaces may delay the biofilm development by some time. The other strategy is to control biofilms by chemical and mechanical cleaning regimes. These often require both a cleaning and disinfecting (sanitizing) for optimal results (Krysinski et al. 1992). This is possible if the surfaces are easily accessible, which is often not the case in industrial processes. There is a need for more research into the development of new strategies and approaches to handling biofilms in industry.

CONCLUSION

Microbial biofilms and biofouling of surfaces and interfaces within industrial environments, including the food and potable water industries, is a major problem. Although these industrial environments contain a diverse set of conditions, they all have problems with biofilms and biofouling on surfaces and require the use of novel detection methodology and control regimes. You will notice that the detection of biofilms within these industries often include the same methods that have been adapted to the individual requirements. The control methods are more diverse, due to restrictions in the types of disinfectants allowed in the different industries. However, they all agree that the strategy is not only to kill the microorganisms on the surface, but also to remove them thus preventing rapid re-growth. This is to delay biofilm formation for as long as possible.

In industry, the first step is identifying the problem of biofilms and biofouling in a particular process or site. Subsequently, it is important to determine the best possible methods for detection of biofilms in situ, so that they can be characterized and possibly further studied in the laboratory. Finally, this information will help to define strategies for controlling biofilm formation in that specific environment. Therefore, the three points of biofilms and biofouling in industry are, identifying the problem, detecting the problem *in situ* and controlling the problem.

REFERENCES

Angles, ML, Marshall, KC and Goodman, AE (1993) Plasmid transfer between marine bacteria in the aqueous phase and biofilms in reactor microcosms. *Applied and Environmental Microbiology* **59**:843–850.

Banks, MK and Bryers, JD (1990) Cryptic growth within a binary microbial culture. *Applied Microbiology and Biotechnology* **33**:596–601.

Bar-Or, Y (1990) The effects of adhesion on survival and growth of microorganisms. *Experientia* **46**:823–826.

Bayer, AE and Lamed, R (1996) Ultrastructure of the cell surface cellulosome of *Clostridium thermocellum* and its interaction with cellulose. *Journal of Bacteriology* **167**:828–836.

Block, JC, Haudidier, K, Paquin, JL, Miazga, J and Levi, Y (1993) Biofilm accumulation in drinking water distribution systems. *Biofouling* **6**:333–343.

Boyd, A and Chakrabarty, AM (1994) Role of alginate lyase in cell detachment of *Pseudomonas aeruginosa*. *Applied and Environmental Microbiology* **60**:2355–2359.

Brading, ME, Jass, J and Lappin-Scott, HM (1996) Dynamics of biofilm formation. In: Microbial biofilms (Eds. Lappin-Scott, HM and Costerton, JW), Cambridge University Press, Cambridge, pp. 46–63.

Bremer, PJ and Geesey, GG (1991) Laboratory-based model of microbiologically induced corrosion of copper. *Applied and Environmental Microbiology* **57**:1956–1962.

Brown, MRW, Allison, DG and Gilbert, P (1988) Resistance of bacterial biofilms to antibiotics: a growth-rate related effect? *Journal of Antimicrobial Chemotherapy* **22**:777–780.

Camper, A, Burr, M, Ellis, B, Butterfield, P and Abernathy, C (1999) Development and structure of drinking water biofilms and techniques for their study. *Journal of Applied Microbiology* **85**:1S-12S.

Camper, AK, Jones, WL and Hayes, JT (1996) Effect of growth conditions and substratum composition on the persistence of coliforms in mixed-population biofilms. *Applied and Environmental Microbiology* **62**:4014–4018.

Carpentier, B and Cerf, O (1993) Biofilms and their consequences, with particular reference to hygiene in the food industry. *Journal of Applied Bacteriology* **75**:499–511.

Challinor, CJ (1991) The monitoring and control of biofouling in industrial cooling water systems. *Biofouling* **4**:253–263.

Charaklis, GW and Marshall, KC (1990) Biofilms: a basis for an interdisciplinary approach. In: Biofilms (Eds. Charaklis, GW and Marshall, KC) Wiley, New York. pp. 3–15.

Costerton, JW, Lappin-Scott, HM and Cheng, K-J (1992) Glycocalyx, Bacterial. In: Encyclopedia of Microbiology, Vol. 2 (Ed. Lederberg, J), Academic Press, Inc. San Diego, pp. 311–317.

Costerton, JW, Lewandowski, Z, deBeer, D, Caldwell, D, Korber, D and James, G (1994) Biofilms, the customized microniche. *Journal of Bacteriology* **176**:2137–2142.

Davies, DG, Parsek, MR, Pearson, JP, Iglewski, BH, Costerton, JW and Greenberg, EP (1998) The involvement of cell-to-cell signals in the development of bacterial biofilm. *Science* **280**:295–298.

de Beer, D, Srinivasan, R and Stewart, PS (1994) Direct measurement of chlorine penetration into biofilms during disinfection. *Applied and Environmental Microbiology* **60**:4339–4344.

Fuqua, WC, Winans, SC and Greenberg, EP (1994) Quorum sensing in bacteria: the LuxR-LuxI family of cell density-responsive transcriptional regulators. *Journal of Bacteriology* **176**:269–275.

Geesey, GG, Mutch, R, Costerton, JW and Green, RB (1978) Sessile bacteria: an important component of the microbial population in small mountain streams. *Limnology and Oceanography* **23**:1214–1223.

Geesey, GG, Iwaoka, T and Griffiths, PR (1987) Characterisation of interfacial phenomena occurring during exposure to a thin copper film to an aqueous suspension of an acidic polysaccharide. *Journal of Colloid Interface Science* **120**:370–376.

Hamilton, WA (1995) Biofilms and microbially influenced corrosion. In: Microbial biofilms (Eds. Lappin-Scott, HM and Costerton, JW), Cambridge University Press, Cambridge, pp. 171–182.

Hausner, M and Wertz, S (1999) High rates of conjugation in bacterial biofilms as determined by quantitative in situ analysis. *Applied and Environmental Microbiology* **65**:3710–3717.

Jass, J and Lappin-Scott, HM (1997) Microbial biofilms in industry: wanted dead or alive? *Chemistry and Industry* **17**:682–685.

Jass, J and Lappin-Scott, HM (1994) Sensitivity testing for antimicrobial agents against biofilms. In: Bacterial biofilms and their control in medicine and industry (Eds. Wimpenny, J, Nichols, WW, Stickler, D and Lappin-Scott, H), Bioline, Cardiff, pp. 73–76.

Keevil, CW and Walker, JT (1992) Normarski DIC microscopy and image analysis of biofilms. *Binary* **4**:93–95.

Keevil, CW, Mackerness, CW and Colbourne, JS (1990) Biocide treatment of biofilms. *International Biodeterioration* **26**:169–179.

Korber, DR, Lawrence, JR, Lappin-Scott, HM and Costerton, JW (1995) Growth of microorganisms on surfaces. In: Microbial biofilms (Eds. Lappin-Scott, HM and Costerton, JW), Cambridge University Press, Cambridge, pp. 15–45.

Korber, DR, James, GA and Costerton, JW (1994) Evaluation of fleroxacin activity against established *Pseudomonas aeruginosa* biofilms. *Applied and Environmental Microbiology* **60**:1663–1669.

Krysinski, EP, Brown, LJ and Marchisello, TJ (1992) Effects of cleaners and sanitizers on *Listeria monocytogenes* attached to product contact surfaces. *Journal of Food Protection* **55**:246–251.

Lappin-Scott, HM and Costerton, JW (1989) Bacterial biofilms and surface fouling. *Biofouling* **1**:323–342.

Lappin-Scott, HM, Costerton, JW and Marrie, TJ (1992) Biofilms and biofouling. In: Encyclopedia of Microbiology, Vol. 1 (Ed. Lederberg, J), Academic Press, Inc. San Diego, pp. 277–284.

Lawrence, JR, Korber, DR, Hoyle, BD, Costerton, JW and Caldwell, DE (1991) Optical sectioning of microbial biofilms. *Journal of Bacteriology* **173**:6558–6567.

Lee, SF, Li, YII and Bowden, GH (1996) Detachment of *Streptococcus mutans* biofilm cells by an endogenous enzymatic activity. *Infection and Immunity* **64**:1035–1038.

Marshall, KC (1996) Adhesion as a strategy for access to nutrients. In: Bacterial adhesion: Molecular and ecological diversity (Ed. Fletcher, M), Wiley-Liss, Inc. New York, pp. 59–87.

Marshall, KC (1994) Microbial adhesion in biotechnological processes. *Current Opinion in Biotechnology* **5**:296–301.

Marshall, KC (1992) Biofilms: an overview of bacterial adhesion, activity and control at surfaces. *ASM News* **58**:202–207.

Mattilla-Sandholm, T and Wirtanen, G (1992) Biofilm formation in the industry: a review. *Food Review International* **8**:573–603.

McLean, RJC, Whieley, M, Stickler, DJ and Fuqua, WC (1997) Evidence of autoinducer activity in naturally occurring biofilms. *FEMS Microbiology Letters* **154**:259–263.

Ophir, T and Gutnick, DL (1994) A role for exopolysaccharides in the protection of microorganisms from desiccation. *Applied and Environmental Microbiology* **60**:740–745.

Pratt, LA and Kolter, R (1998) Genetic analysis of *Escherichia coli* biofilm formation: roles of flagella, motility, chemotaxis and type I pili. *Molecular Microbiology* **30**:285–293.

Roberson, EB and Firestone, MK (1992) Relationship between desiccation and exopolysaccharide production in a soil *Pseudomonas* sp. *Applied and Environmental Microbiology* **58**:1284–1291.

Solano, C, Sesma, B, Alvarez, M, Humphrey, TJ, Thorns, CJ and Gamazo, C (1998) Discrimination of strains of *Salmonella enteritidis* with differing levels of virulence by an in vitro glass adherence test. *Journal of Clinical Microbiology* **36**:674–678.

Stewart, PS, Griebe, T, Srinivasan, R, Chen, C-I, Yu, FP, deBeer, D and McFeters, GA (1994) Comparison of respiratory activity and culturability during monochloramine disinfection of binary population biofilms. *Applied and Environmental Microbiology* **60**:1690–1692.

Stewart, PS, Peyton, BM, Drury, WJ and Murga, R (1993) Quantitative observations of heterogeneity's in *Pseudomonas aeruginosa* biofilms. *Applied and Environmental Microbiology* **59**:327–329.

Stoodley, P, Dodds, I, Boyle, JD and Lappin-Scott, HM (1999) Influence of hydrodynamics and nutrients on biofilm structure. *Journal of Applied Microbiology* **85**:19S–28S.

Stoodley, P, deBeer, D and Lewandowski, Z (1994) Liquid flow in biofilm systems. *Applied and Environmental Microbiology* **60**:2711–2716.

Suci, PA, Mittelman, MW, Yu, FP and Geesey, GG (1994) Investigation of ciprofloxacin penetration into *Pseudomonas aeruginosa* biofilms. *Antimicrobial Agents and Chemotherapy* **38**:2125–2133.

Surette, MG and Bassler, BL (1998) Quorum sensing in *Escherichia coli* and *Salmonella typhimurium*. *Proceedings of the National Academy of Sciences* **95**:7046–7050.

Väisänen, OM, Nurmiaho-Lassila, E-L, Marmo, SA and Salkinoja-Salonen, MS (1994) Structure and composition of biological slime on paper and board machines. *Applied and Environmental Microbiology* **60**:641–653.

Vandevivere, P and Kirchman, DL (1993) Attachment stimulates exopolysaccharide synthesis by a bacterium. *Applied and Environmental Microbiology* **59**:3280–3286.

Walker, JT, Dowsett, AB, Dennis, PJL and Keevil, CW (1991) Continuous culture studies of biofilm associated with copper corrosion. *International Biodeterioration* **27**:121–134.

Walker, JT, Rogers, J and Keevil, CW (1994) An investigation of the efficacy of a bromine containing biocide on aquatic consortium of planktonic and biofilm microorganisms including *Legionella pneumophila*. *Biofouling* **8**:47–54.

Wimpenny, JWT, Kinniment, SL and Scourfield, MA (1993) The physiology and biochemistry of biofilm. In: Microbial biofilms: formation and control (Eds. Denyer, SP, Gorman, SP and Sussman, M), Blackwell Scientific Publ., Oxford, pp. 51–94.

2 Biofouling in Drinking Water Systems

2.1 Problems of Biofouling in Drinking Water Systems

ANNE K. CAMPER[1] and GORDON A. McFETERS[2]
[1]*Center for Biofilm Engineering and Department of Civil Engineering*
[2]*Department of Microbiology and Center for Biofilm Engineering,*
Montana State University, Bozeman, MT 59717, USA

INTRODUCTION

At first glance, it would not seem that conditions in a drinking water distribution system are conducive to microbial growth. Treatment is optimized to remove particles, including bacteria. Temperatures are generally low, in most cases there is little organic carbon present, and disinfectants are deliberately added to suppress microbial growth. Remarkably, microbial growth has been found in all distribution systems examined, including stainless steel pipes delivering ultrapure water.

Biofilms in drinking water became important with the recognition that coliform proliferation on surfaces may result in coliform positive water samples. This is of considerable regulatory concern, since these organisms are used as indicators of the microbiological safety of potable water. In addition to the regrowth issue, biofilms in distribution systems can cause other negative effects on finished water quality. *Actinomycetes* or fungi may result in taste and odour problems (Olson 1982; Burman 1965, 1973), which then lead to consumer complaints. Bacterial biofilms may contribute to the corrosion of pipe surfaces (Lee *et al.* 1980). Iron bacteria can grow on ferrous metal surfaces (Ridgway *et al.* 1981) and particulate iron may be released into the water (Victoreen 1974). In this case, red water may be the outcome. It is also possible to have nitrifying bacteria present in biofilms and these organisms could result in nitrification episodes in systems where chloramines are used (Wolfe *et al.* 1990).

REGROWTH IN DISTRIBUTION SYSTEMS

Coliforms are used as the surrogate, or indicator, of pathogen presence. By definition, an indicator organism should be present only in the event of

Industrial Biofouling. Edited by J. Walker, S. Surman, J. Jass
©2000 John Wiley & Sons Ltd

faecal contamination, survive longer than a pathogen, and be present in higher concentrations than the pathogen. The presence of the coliforms is interpreted to represent recent post-treatment faecal pollution (Pipes 1990). One major exception has been accepted in the case of regrowth, where the condition of recent faecal contamination is exempted. It is known that some coliforms can persist in, and sometimes regrow in biofilms of both distribution systems and pipe pilot systems (Camper 1994), with cells detaching and contaminating the bulk water phase. Coliforms released from biofilms may result in elevated coliform positive water samples even though system integrity and disinfectant residual have been maintained (Smith *et al.* 1990; Characklis 1988; Haudidier *et al.* 1988). In these cases, recent faecal contamination is not indicated and a 'false positive' is obtained. If a utility in the U.S. can prove that the presence of coliforms is attributed to a biofilm regrowth event, it can apply for variance from the Total Coliform Rule. Past experience has indicated that this may be reasonable, since no instances of waterborne outbreaks have been associated with regrowth events. However, the opportunistic pathogens, *Klebsiella* spp. and *Aeromonas* spp., are known to be components of biofilms and associated with regrowth events (Geldreich and Rice 1987; Camper 1996; van der Kooij 1988; Havelaar *et al.* 1990).

Regrowth of coliforms and heterotrophs has been attributed to a host of variables including: (i) temperature effects, especially warm water conditions; (ii) the amount of utilizable carbon for substrate; (iii) inefficiencies in the removal/disinfection of organisms in treatment; (iv) the presence of corrosion products in distribution systems; (v) disinfectant dose/type; and (vi) distribution system hydrodynamics (Smith *et al.* 1990). These topics have been discussed in review articles by Camper (1994), LeChevallier (1990) and Block (1992). A critical variable is the presence of a disinfectant. Intuitively, elevated levels of chlorine should control regrowth, but this is often not the case (LeChevallier *et al.* 1990; Martin *et al.* 1982; Reilly and Kippen 1984; Oliveri *et al.* 1985; Centers for Disease Control 1985; Hudson *et al.* 1983). There is the potential for these same conditions to influence the long-term persistence of pathogenic organisms in biofilms on distribution system pipe surfaces.

PUBLIC HEALTH ISSUES

In the years 1993–1994, the U.S. had 30 disease outbreaks associated with water. Of these, 20 were from groundwater supplies. Ten of the 25 outbreaks of gastroenteritis for which the aetiological agent was identified were caused by protozoa, eight by chemical poisoning, three by *Campylobacter jejuni*, two by *Shigella* spp., and one each by *Salmonella typhimurium* and a non-O1 *Vibrio cholerae* (Kramer *et al.* 1996). Data from this survey are given to indicate that waterborne disease is still present and that the etiological agents are varied. It is also critical to note that in half of the waterborne outbreaks in the United States, no aetiological agent is identified (Craun *et al.* 1997). It is probable that many of these cases are the result of the inability to detect and culture various viruses responsible for waterborne disease, as well as the difficulties in culturing and identifying bacteriological agents. It is unknown if any of these outbreaks are specifically the result of biofilms in distribution or treatment systems.

The evaluation of water associated disease outbreaks such as hospital acquired (nosocomial) respiratory infections by organisms in water (*Legionella, Pseudomonas, Mycobacterium*) or other mild gastroenteritis (*Aeromonas*) are not included in typical drinking water surveys. This is because the ability to associate disease caused by some of these organisms with water is difficult. In fact, an overview of risk assessment on health effects from microbes shows that there are many organisms responsible for disease and that there are significant problems associated with establishing risk (Sobsey *et al.* 1993). With this overview in mind, the following sections on the organisms perceived and known to be responsible for waterborne and water associated disease and their potential association with biofilms have been prepared. It should be noted that there appears to be the potential for a direct link with biofilms in several instances.

BACTERIA

General Heterotrophs

There are three 'categories' of bacteria of potential public health concern that can be isolated from drinking water. The first are the general heterotrophic bacteria. In the past, these bacteria have not been considered to be of significance, but their presence in finished water suggests that the water is capable of supporting microbial growth. With the growing number of immunocompromised people in the population, it is probable that interest in the number of heterotrophs present in drinking water will increase. The studies of Payment *et al.* (1991; 1993) indicated that populations drinking conventionally distributed water may be at higher risk for gastroenteritis than those drinking the same water after reverse osmosis treatment. There was no correlation of illness with coliform levels, but there was an association with the increased number of heterotrophs enumerated at 35° C, as seen with longer hydraulic residence times. The correlation was attributed to regrowth of the heterotrophs rather than contamination of the distribution system.

Opportunistic Pathogens

The second set of organisms are the opportunistic pathogens. These organisms do not cause disease in healthy individuals, but have been implicated with health effects in the portion of the population with lower immune responses (the very young, very old, and those with immuno-deficiencies due to chemotherapy, therapy for implants or organ replace-ment and HIV infection). A review of opportunistic pathogens in drinking water has been prepared by Geldreich (1991).

The *Aeromonas* spp. are an example of a group of opportunistic pathogens that have received scrutiny in the past few years. A review of the information available on this genus reveals a general lack of understanding of how these bacteria behave in the aquatic environment and also demonstrates the controversy surrounding their public health significance.

Aeromonas spp. have been isolated from drinking water (Versteegh *et al.* 1989; Burke *et al.* 1984b; LeChevallier *et al.* 1982; Havelaar *et al.* 1990; van der Kooij 1988). They are considered water-based rather than waterborne organisms, since they are indigenous to aquatic environments and physiologically adapted for growth within the drinking water system (Rippey and Cabelli 1979). Aeromonads appear to be a small and variable component of the overall heterotrophic population (Havelaar *et al.* 1990), with conflicting evidence supporting the correlation of aeromonads with the overall heterotrophic plate count population (Havelaar *et al.* 1990; LeChevallier *et al.* 1982). They can use a wide range of biopolymers; this may be important for aeromonads to maintain viability or grow in a distribution system, particularly if the aeromonads are growing in the biofilm and using products from more predominant bacteria (van der Kooij 1988). It has been shown that the presence of other microflora, such as pseudomonads, in bottled water enhanced the survival of *Aeromonas* (Warburton *et al.* 1994).

The factors governing aeromonas regrowth appear highly complex and are not well understood. In some studies, positive correlation has been made with temperature and residence time (Havelaar *et al.* 1990), although temperature has been shown to have no correlation in other investigations (Burke *et al.* 1984b). However, these studies collectively agree that aeromonas numbers have no correlation to coliform levels. Colbourne *et al.* (1991) demonstrated that *Aeromonas hydrophila* remained as a component of the biofilm even when the monochloramine con-centration was 0.6 mg L^{-1}.

The potential for waterborne aeromonas to act as aetiological agents in a disease outbreak is also poorly understood. It has been shown (Burke *et al.* 1984a, b) that aeromonas numbers in the distribution system correlated with the numbers of clinical gastroenteritis cases associated with aeromonas. In a survey of studies investigating the presence and absence of aeromonas in faeces, particularly in patients suffering from diarrhoea (van der Kooij 1988), it was found that isolation rates varied widely (<1 to 20%), with highest isolation frequencies in tropical regions and lowest in Europe and the USA. The most common isolate was *A. caviae*. He also noted that the isolation of aeromonas in the absence of other pathogens is not adequate evidence of it being the aetiologic agent.

Concerns associated with aeromonas have led to some attempts at regulation of these organisms. For example, health authorities in the Netherlands have defined 20 colony-forming units (CFU) 100 mL^{-1} in drinking water at the production plant and 200 CFU 100 mL^{-1} during distribution, as maximum allowable values (van der Kooij 1988). In Canada, a limit of 0 CFU 100 mL^{-1} in bottled water has been proposed (Warburton *et al.* 1994).

Overall, it can be concluded that while *Aeromonas* demonstrably has a niche in distribution system microbiota and possesses the potential for causing waterborne disease, direct cause and effect between the presence of the bacteria in water and a disease outbreak is questionable.

Other opportunistic pathogens found to be capable of proliferating in drinking water distribution systems include *Mycobacterium* spp. (Engel *et al.* 1980; Kaustova *et al.* 1981; Collins *et al.* 1984), *Legionella* spp. (Dennis *et al.* 1982; Tobin *et al.* 1981b; Wadowsky *et al.* 1982; Rogers and Keevil 1992; Colbourne *et al.* 1988) and *Pseudomonas* (Geldreich 1990; Gambassini *et al.* 1990).

The mycobacteria are found in drinking water distribution systems throughout the United States (duMoulin *et al.* 1988; Fischeder *et al.* 1991; von Reyn *et al.* 1993; Glover *et al.* 1994). Specific isolates have been shown to survive for up to 41 months in a distribution system (von Reyn *et al.* 1994). The organisms infect the lungs and may cause a tuberculosis-type disease. Of particular interest are the organisms belonging to the *M. avium* complex (*M. avium* and *M. intracellulare*; MAC). These are acid-fast rod shaped bacteria that are found in natural waters, including drinking water, throughout the United States (Haas *et al.* 1983; duMoulin *et al.* 1986, 1988; Carson *et al.* 1988; Fischeder *et al.* 1991; von Reyn *et al.* 1993, 1994; Glover *et al.* 1994). It appears that these organisms are actually growing in these environments, since they have been shown to reproduce in water with no added nutrients (von Reyn *et al.* 1994). Several investigators have recovered these organisms from hospital water systems (Carson *et al.* 1988; duMoulin *et al.* 1988; von Reyn *et al.* 1993) in numbers ranging from 1–10 000 CFU per 100 ml water. The source of these bacteria is believed to be shed biofilm fragments. This is substantiated by the observation by Schultz-Robbecke *et al.* (1989, 1992) that 45 of 50 biofilms removed from municipal or domestic water sources contained mycobacteria. Another adaptive mechanism for survival in drinking water is their ability to tolerate disinfection by chlorine (Haas *et al.* 1983; Carson *et al.* 1978; Collins *et al.* 1984). A recently completed study found that *Mycobacterium* spp. were present in five of eight biofilm samples grown in conventionally treated water. These numbers could be reduced if the water was depleted in organic carbon or if a free chlorine residual was maintained (LeChevallier *et al.* 1998).

The MAC is of sufficient public health concern that they are one of the top three bacterial 'emerging pathogens' in the drinking water industry. Consequently, they have been added to the U.S. Environmental Protection Agency's (EPA) Contaminant Candidate list, which identifies organisms of concern worthy of further research for risk assessment. These organisms are also candidates for possible regulation by the EPA.

In distribution systems, *Legionella* is believed to be most prevalent in biofilms, since it depends on other organisms for growth-supporting substances (Wadowsy and Yee 1983, 1985; Stout *et al.* 1985; Rogers *et al.* 1994). These organisms colonize and grow in water heaters, shower heads and cooling towers, where their release can lead to respiratory disease in sensitive individuals. A review paper by Lin *et al.* (1998) suggests that hospitals take routine samples for the organism in their distribution system and the efficacy of any disinfection processes be monitored before and after analysis for the presence of *Legionella*.

The pseudomonads are ubiquitous in nature. In most cases they are benign, but have been associated with infections in burn patients, eye infections, and in the special case of cystic fibrosis.

Frank Pathogens

The third subset of bacteria are the frank pathogens. These are organisms known to be associated with waterborne disease outbreaks and include the genera *Salmonella*, *Shigella*, *Escherichia*, *Campylobacter* and *Vibrio*. It is generally believed that these organisms do not last long outside of their hosts and are easily disinfected. To colonize and persist in distribution systems, the pathogens must be able to successfully compete with the heterotrophic bacterial populations. Competition with the existing micro-flora may be a key parameter in preventing proliferation of pathogenic organisms. In research using laboratory columns containing granular activated carbon (GAC), Camper *et al.* (1985) found that a suite of pathogenic bacteria could survive on GAC when fed a sterile source of surface water. However, if the pathogens were challenged with organisms present in unsterilized surface water, the numbers of pathogens declined. If pathogens were added to the carbon simultaneously with the autochtho-nous heterotrophs, they declined more rapidly than in the first instance. Finally, if pathogens were added to previously colonized GAC, the decline of pathogens in the filter and filter effluent was the most rapid. In recently completed experiments using groundwater and laboratory columns with virgin GAC, GAC that had been in operation in a filter for a few months and biologically activated carbon (BAC) from a full-scale plant, coliform elimination was the most rapid from the BAC filter (LeChevallier et al. 1998). Other laboratory studies (Rollinger and Dott 1987) with several pathogens on GAC reported similar results. The pathogens persisted when introduced to sterile GAC and fed sterile water, but were eliminated from the medium when they were subsequently challenged by autochthonous bacteria from tap water. It should be noted that GAC filter effluents have been found to contain GAC particles colonized with a variety of potential pathogens including Escherichia coli, Klebsiella pneumoniae and Aeromonas hydrophila (Brewer and Carmichael 1979; Tobin et al. 1981a; Camper et al. 1986).

Salmonella is well recognized as a pathogenic organism of faecal origin. It therefore should not be present in a well-run potable water distribution system. If found in water, it is regarded as more a case of survival rather than proliferation of the organism.

With the exception of the very large outbreak in Riverside, California, there seems to be a level of about 1–100 cases per year of waterborne salmonellosis in the U.S. (Craun 1986). There is considerable potential for non-detection, given phenomena such as viable non culturable (VBNC) cells with Salmonella. In an investigation with Salmonella enteritidis in river water microcosms (Roszak et al. 1983), salmonellae rapidly became non culturable. This was initially reversible following nutrient addition, but after 3 weeks resuscitation failed to give culturable cells, although the cells remained viable. Therefore, even if water during a disease outbreak tests negative for salmonellae, there remains the possibility of transmission by VBNC cells.

The importance of attachment to environmental surfaces in the survival of this organism has been previously investigated. Camper et al. (1985) found Salmonella readily colonized and persisted on GAC in water, although attachment was at a lower rate and the organism decreased in numbers more rapidly in the presence of other heterotrophic bacteria. Suspended pathogenic cells died away faster than attached cells. We have completed experiments where S. typhimurium was able to colonize simulated drinking water distribution system biofilms and remain detectable by fluorescent antibodies for up to two months. Some of these organisms could be cultured after recovery in a non-selective medium (Warnecke 1996). However, there is no information available on the virulence of these bacteria.

Some strains of E. coli such as the enterohemorrhagic O157:H7 strain (ECO157) have a demonstrated pathogenicity to humans. This strain has several physiological differences from typical isolates, including a lack of the enzyme β-glucuronidase (often used for *E. coli* detection) and poor or no growth at 45° C (Rice *et al.* 1992). Therefore, this strain is non-detectable by standard methods used for the routine detection of *E. coli*. It has been shown to persist at a similar rate as typical *E. coli* strains under drinking water conditions, confirming the premise that typical *E. coli* strains would be effective in indicating ECO157 presence (Rice *et al.* 1992). However, this study did not account for differential survival on particulates or in biofilms. Survival on surfaces has been shown to be the major long-term persistence mechanism of *E. coli* in lake waters (Brettar and Höfle 1992). Pathogenic and non-pathogenic *E. coli* strains have been shown to have similar growth rates to an environmental isolate of *E. coli* under growth conditions relevant to drinking water distribution systems (Camper *et al.* 1991). Colonization by an environmental *E. coli* isolate of a pre-existing biofilm has been demonstrated under water distribution system conditions (Robinson *et al.* 1995). In addition, this organism was shown to persist in a model distribution system biofilm for over 12 days (Block 1992) and for at least 21 days in another system even in the presence of 0.3 mg L^{-1} monochloramine (Colbourne *et al.* 1991). A faecal origin, non-benzoate degrading *E. coli* has been shown to be able to colonize a reactor containing a biofilm of benzoate degrading bacteria and subsequently re-enter the water phase. In this study, 5 mM benzoate was the sole carbon source, demonstrating consortial feeding by this organism (Szewzyk *et al.* 1994). The same investigator had also demonstrated that a pathogenic strain of *E. coli* could colonize a single species biofilm (Szewzyk and Manz 1992). Studies in our laboratories have shown that *E. coli* O157:H7 did not persist for over one week in a mixed-population biofilm grown under drinking water conditions (Warnecke 1996).

VIRUSES AND PROTOZOANS (*GIARDIA* AND *CRYPTOSPORIDIUM*)

These two types of pathogens have been grouped by the characteristic that they do not have the ability to grow in the distribution system. In the case of bacteria, there is always the chance that a pathogen can colonize and grow in a biofilm, while this is not possible for the pathogenic viruses and the specific protozoans listed above. Virus and cyst/oocyst interactions with biofilms would be more similar to that of particles.

There are several groups of viruses believed to be of importance in waterborne outbreaks, although no information exists in the open literature as to their potential presence or survival in biofilms. The enteroviruses, including those that cause polio, can be very resistant to disinfection and there have been unsubstantiated reports that a polio outbreak was attributed to a drinking water source (Mosley 1966). More information on these viruses in potable water is being collected under the Information Collection Rule in the United States. Hepatitis A virus (HAV) can survive for more than four months in water (Sobsey *et al.* 1988) and has been identified as the causative agent in more than 20% of waterborne disease outbreaks in the U.S. where the aetiological agent has been determined (Lippy and Waltrip 1984). However, it should be noted that many of these outbreaks were associated with groundwater rather than surface water sources. Although the virus is sensitive to disinfectants, it has been found in samples with a free chlorine residual of 0.2 mg/L (Bosch *et al.* 1991). Again, no direct evidence for a biofilm/virus interaction presently exists.

Giardia and *Cryptosporidium* cysts and oocysts are notoriously resistant to disinfection and can survive for extended periods of time in cold waters. As in the case for the viruses, there is no documented evidence that these organisms have accumulated in distribution systems and subsequently caused disease. There is some evidence suggesting that protozoan cysts and oocysts can associate and remain in laboratory grown biofilms for several months, but there is no evidence that these organisms are still virulent (Rogers and Keevil 1995). Practical experience indicates that once the water carrying the cysts/oocysts starts to leave the system, there are decreasing instances of disease. There are no reported instances where there has been a reoccurrence of a protozoan disease outbreak in the same distribution system once the original source of the organism was removed or corrected. This may provide indirect evidence that the cysts (i) do not accumulate in numbers sufficient to cause disease if biofilm is sloughed from the surface or (ii) lose infectivity and virulence with time even if they do accumulate.

MANAGEMENT OF BIOFILMS IN DISTRIBUTION SYSTEMS

Practical experience has shown that there are a variety of water quality parameters that tend to support biofilm growth in drinking water systems. These include temperature, the concentration and type of organic carbon and disinfectant and the presence or absence of a corrosion control regime when corrodible materials are used in the distribution system. Many of these interactions have been supported in laboratory studies or at the pilot scale using experimental conditions similar to those in full-scale distribution systems. However, there are still conflicting and confusing aspects to this work. Some researchers have shown that the concentration of organic carbon measured as assimilable organic carbon (AOC) or biodegradable organic carbon (BDOC) influences biofilm cell numbers, while others have less conclusive results. In fact, biofilms are known to occur in ultrapure water stainless steel distribution systems where organic levels should be extremely low. There are also conflicting results on the ability of chlorine or monochloramine to limit biofilm proliferation. In many cases, the findings appear to be system specific and indicate the complexity of the response of the biofilm to an interrelated set of environmental conditions. One general observation can be made; biofilms will not be eliminated from distribution systems by any of the current methods available now or in the future. Our primary challenge is to control rather than eradicate biofilms from distribution systems.

Reducing Organic Levels

Reducing organic carbon concentrations in treatment is gaining favour as an acceptable option for reducing biofilm growth, controlling the formation of disinfection by-products and reducing disinfectant demand. As regulations become increasingly more stringent, the relative economic gains from organic removal are increasing.

There are a variety of means for reducing organic matter in drinking water. Although it is beyond the scope of this chapter to provide detailed descriptions of the processes, a list of technologies is provided. These include changing the source water, enhanced coagulation, activated carbon adsorption, membrane filtration and biological filtration. The choices available to a specific utility will be limited by water quality and physical and economic resources. Regardless of the type of treatment, a reduction in the amount of organic carbon can potentially reduce the amount of biofilm development in the distribution system.

Attempts have been made to define potential threshold concentrations of organic carbon that limit biofilm growth. van der Kooij (1992) has suggested an AOC level of 10 μg C l^{-1} for heterotrophs, while LeChevallier et al. (1991) recommended a level of 50 μg C l^{-1} for coliform control. In terms of BDOC, Servais et al. (1991) have associated biological stability of water with a level of 0.2 mg l^{-1}, although, Joret (1994) has stated these levels are temperature dependent (0.15 mg l^{-1} at 20° C and 0.30 mg l^{-1} at 15° C). These numbers have been used with some success as design criteria in European systems where the confounding influence of high levels of secondary disinfectant is not present. There are instances where there has not been a clear association between AOC/BDOC and biofilm development. For example, Kerneis et al. (1994) reported that there was no correlation between distribution system BDOC measurements, suspended bacteria and fixed biomass. Conflicting information was also obtained in field studies; regrowth was seen in systems with average AOC levels both greater than and less than 100 μg L^{-1} (LeChevallier et al. 1996a). In experiments where we have used mild steel pipe loops and annular reactors there was a weak correlation between biofilm and influent AOC concentrations, but no correlation with the concentration of AOC in the reactors (Camper 1996). These results suggest that utilizable organic matter as a sole parameter in determining the potential for a water to support biofilm is not always appropriate.

Optimizing Disinfection

An explanation for the lack of consistent correlation between utilizable organic matter and biofilm formation is the presence of a disinfectant. There appear to be interactions between the organic matter, the pipe surface and the biofilm that influence the number of organisms present. It is generally believed that increasing the concentration of a disinfectant should control regrowth, but many instances exist where the opposite effect is seen (LeChevallier et al. 1987; Martin et al. 1982; Reilly and Kippen 1984; Oliveri et al. 1985; CDC 1985; Hudson et al. 1983). When distribution system biofilms were examined, no correlation was found between free chlorine residuals and the number of heterotrophic plate count (HPC) organisms per unit surface area (Hudson et al. 1983). Many reports exist that demonstrate the relative lack of sensitivity of biofilm cells to disinfectants. This effect is even more pronounced if the biofilms are grown on reactive iron surfaces (Kerneis et al. 1994; LeChevallier et al. 1990; Chen et al. 1993). It has been noted that increased corrosion rates decrease the efficacy of free chlorine against biofilm organisms (LeChevallier et al. 1993).

As a result of changing drinking water regulations and an increased emphasis on the presence of disinfection by-products, utilities may use monochloramine as a secondary disinfectant. In the past, monochloramine was not viewed favourably because of its high CT requirement as compared to chlorine, but an interesting beneficial effect on biofilms was noted. In a distribution system comparison, chloramines were more effective at reducing the number of biofilm total coliforms and HPC than chlorine (Neden et al. 1992). In another study where statistical evaluation of the influence of chloramine concentration on attached microbial populations in a distribution system was made, an inverse relationship was established (Donlan and Pipes 1988). A study in a model pipe loop system composed of several materials showed that biofilms on galvanized steel, copper, or polyvinylchloride (PVC) surfaces were readily disinfected by free chlorine or monochloramine (1 mg l^{-1}), while iron pipe surface-associated bacteria were more susceptible to monochloramine than free chlorine (4 mg l^{-1}) (LeChevallier et al. 1990). A field study of 31 utilities showed systems that used chloramines had 0.51% coliform positives in 35 159 water samples as compared to 0.97% of 33 196 samples in chlorinated systems. These same data showed that the average density of coliforms in the water of the chlorinated systems was 35 times higher than in the chloraminated systems (LeChevallier et al. 1996b).

Corrosion Control

An interesting recent development in biofilm management occurred when utilities implemented corrosion control to comply with the Lead and Copper Rule. Anecdotal reports both of decreased bacterial numbers and increased disinfectant residual have been given, especially if the distribution system has a substantial amount of unlined iron-containing pipe. Iron pipes have been implicated as a key component in microbial regrowth in distribution systems (Camper 1996; Camper et al. 1996; LeChevallier et al. 1996b; LeChevallier et al. 1993). This has been supported by observations that utilities with a large proportion of unlined ferrous metal pipes are more prone to coliform regrowth. A utility survey has also shown a positive relationship between the number of miles of unlined metal pipes and coliform occurrences (LeChevallier, et al. 1996b). Pilot system experiments have supported the interaction between corroding iron pipes and biofilms. Neden et al. (1992) found that bacterial populations on unlined cast iron were the highest, while PVC was colonized with the lowest number and Block (1992) determined that there was a progressive decrease in bacterial densities on surfaces from cast iron, tinned iron, cement lined cast, to stainless steel. The influence of iron pipes can be quite dramatic, with one experiment demonstrating a >100 fold increase in biofilm cell numbers on iron compared to PVC surfaces (LeChevallier et al. 1998). It has also been demonstrated that biofilms on ferrous metal surfaces were less susceptible to free chlorine than biofilms on inert materials, (LeChevallier et al. 1987, 1988, 1990), even in the presence of measurable chlorine residuals (LeChevallier et al. 1998). This may be because the metal exerts a chlorine demand. The ability for reduced iron to react with disinfectants has been documented (Knocke 1988; Knocke et al. 1994; Vasconcelos et al. 1996). Research in our laboratories has confirmed that pipe material has a dramatic impact on biofilm cell numbers. Mild steel surfaces were consistently colonized by nearly ten-fold more heterotrophs and two to ten-fold more coliforms than polycarbonate. The effect was also seen in effluent cell counts. Interestingly, the presence of a small amount of mild steel (10% on the basis of surface area) in an otherwise polycarbonate reactor resulted in elevated biofilm counts on all surfaces; the plastic surfaces supported the same numbers of bacteria as seen on the steel itself (Camper et al. 1996).

As stated above, there is evidence to demonstrate that corrosion control has mitigated coliform regrowth in full scale systems (Hudson et al. 1983; Lowther and Moser 1984; Martin et al. 1982; Schreppel et al.

1997). It is unknown if the reason is improved disinfection, reduced attachment sites for bacteria or other factors. LeChevallier et al. (1993) showed that corrosion control reduced biofilm numbers but attributed the response to increased chlorine efficacy due to decreased corrosion rates. This may not be the case if the corrosion control scheme is an increase in pH, since chlorine speciation and disinfection is adversely influenced by elevated pH. However, Martin et al. (1982) showed that increasing the pH to 9 in a chlorinated system actually reduced bacterial counts. In this case it may be inferred that corrosion control superseded the effects of reduced disinfection efficacy. We have noted that at near neutral pH in the absence of corrosion control, the presence of low levels of disinfectant actually increases biofilm density on ductile or steel surfaces. This is presumably because corrosion was enhanced and the disinfectant consumed at the surface (Camper 1996; Abernathy 1998). Recently completed laboratory and pilot work has shown that the number of biofilm bacteria is directly related to the mass of corrosion products present; reduction in biofilm and corrosion product accumulation can be achieved by corrosion control schemes or by changing from chlorine to monochloramine to produce less corrosion (Abernathy 1998). In another pilot experiment, the results indicated that corrosion control was more important in reducing bacterial numbers in biofilms than decreasing the amount of BDOC in the water or the maintenance of a free chlorine residual (LeChevallier et al. 1998).

MONITORING AND TESTING FOR BIOFILMS IN DISTRIBUTION SYSTEMS

In general, access to full-scale distribution systems for biofilm samples is extremely limited. As a consequence, it is unlikely that a utility can obtain much information on biofouling directly from pipe sections. Because of this limitation, the tendency is to infer biofilm responses based on easily obtained water samples. There are problems with this approach. If a disinfectant is present, the detached biofilm cells in the water may be reduced, giving the false indication that the biofilms are also controlled. There are no acceptable methods for directly correlating numbers of bacteria in the water with the number in biofilms. Because of these limitations, water distribution personnel may wish to use a side-stream or pilot device to obtain biofilm samples. It should be noted that these devices are appropriate for providing data on trends in biofouling, but it is not appropriate to infer actual numbers of cells in distribution systems from those obtained in side-stream or pilot devices.

Monitoring Devices; Pipe Loops and Reactors

Pilot-scale distribution facilities are typically constructed to provide a platform for determining the influence of water quality parameters on full-scale distribution system performance. This is required when access to the full scale system is limited, when the parameters to be tested may significantly alter the quality of water delivered to the consumer, or if more tightly controlled conditions are required (flow, water quality, pipe composition, etc.). Since pilot system results are extrapolated to the distribution system, it is critical that the reactors be designed to simulate relevant operational conditions, including shear stress, pipe materials, temperature, disinfectant types and concentrations and other water quality parameters such as organic carbon concentrations. Coupons for sampling surfaces should be flush mounted to reduce perturbations in the local hydrodynamics. Ideally, the reactors should be relatively compact, easily controlled, have minimal water demand and be inexpensive to construct, maintain and operate. An additional consideration is the manner in which the pilot system can be mathematically modelled, i.e. either plug flow or completely mixed reactor behaviour.

Historically, pilot systems have been designed to provide information on corrosion (Levin and Schock 1991; Gardels and Sorg 1989; Heumann 1989; Birden *et al.* 1985; AWWA 1985) and more recently to investigate biological processes in distribution systems (LeChevallier *et al.* 1990; Haudidier *et al.* 1988; Camper 1996). For example, the pipe loop at Nancy in France was designed so that experiments to examine biofilm accumulation and detachment of bacteria, disinfectant decay and efficacy against biofilm and suspended organisms and disinfection by-product formation are possible (Haudidier *et al.* 1988).

As an alternative to pipe loops, Characklis (1988) and van der Wende and Characklis (1990) utilized annular reactors in series to simulate the hydraulic conditions found in a distribution system. Annular reactors consist of a stationary outer cylinder with flush mounted coupons for surface analyses and a rotating inner cylinder with an annular space occupied by the water. Rotation of the inner drum controls shear stress, which can be scaled to that found in a circular pipe. Annular reactors are also well-mixed, and when staged, the entire series can simulate long residence times typical of municipal systems.

These reactors have been used by a number of utilities as side-stream devices in treatment plants and distribution systems and to provide advance information about proposed changes in treatment or disinfection practices. For example, a utility may choose to pilot corrosion control in two reactors. One would receive water as it is currently treated and the other would receive water amended with the proposed corrosion control scheme.

Direct comparisons on biofouling between the two systems could then be obtained.

Researchers at the University of Nancy in France have developed a reactor called the Propella® with design criteria somewhat similar to that of the annular reactor. The Propella® has a stationary outer cylinder made of pipe material with removable coupons. There is an inner cylinder with a propeller at the top that forces water down through the inner cylinder, out the bottom and along the inside of the outer cylinder to simulate flow conditions in a distribution system.

Analysis of Biofilm Samples

Culturing

Traditionally, environmental bacteria have been studied by culture-dependent methods. For biofilm samples, the cells are scraped from the surface, dispersed by methods such as homogenization and the bacteria enumerated on a variety of media. Optimal results for heterotrophic bacteria can be obtained by using the spread plate method on R2A agar incubated for one week at room temperature. Other culturing methods can be used if specific subsets of bacteria are being targeted.

It is now well established that cultural methods underestimate the numbers and diversity of environmental bacteria. For example, Wagner *et al.* (1993) found that viable plate counts of bacteria from activated sludge were about 1% of direct microscopic counts. Regardless of the shortcomings of culturing methods, they still have their value. They have regulatory significance since most contaminant levels are based on culturable counts. There are some types of samples, including those with very high concentrations of corrosion products and detritus, that are not amenable to any other type of enumeration method currently available. In this case, culturing using optimal recovery methods provides the best estimate of bacterial numbers.

Nucleic Acid Stains and Physiological Indicators

If clean samples are available, direct microscopy with a variety of nucleic acid and physiological probes can be used to assess the number and activity of biofilm cells. These stains can be used on intact or dispersed (scraped and homogenized) biofilm samples.

Overviews of the use of fluorochrome staining and direct microscopic observation of bacterial cells have been prepared by McFeters *et al.* (1995) and Kepner and Pratt (1994). They and others have noted the general use of the nucleic acid stains acridine orange (AO) and 4′,6-diamidino-2-phenylindole

(DAPI) for obtaining total cells counts. Physiological stains vary from those that measure membrane integrity to those that detect specific metabolic functions. Each fluorochrome has its advantages, although it is recommended that a suite of fluorochromes be used to assess overall activity. For example, Yu *et al.* (1994) reported that 5-cyano-2,3-ditolyl tetrazolium chloride (CTC) and rhodamine 123 (Rh 123) were effective indicators of metabolically active cells in biofilms, but Morin and Camper (1997) reported a lack of sensitivity with CTC in chlorinated biofilms.

Nucleic acid stains and physiological fluorochromes do not identify species but may be used in conjunction with specific antibodies or nucleic acid probes, provided the emission spectra of the different fluorochromes are dissimilar enough to allow separate detection of each component. Hicks *et al.* (1992) combined DAPI for total cell counts with a specific oligonucleotide probe labelled with tetramethyl rhodamine isothiocyanate (TRITC). Pyle *et al.* (1995) used CTC and a fluorescent antibody for *E. coli* O157:H7 in the same assay to detect actively respiring cells.

Fluorescent Antibodies

In specific instances, fluorescent antibodies have been shown to be effective at targeting bacterial cells in intact or dispersed biofilms. Rogers and Keevil (1992) have used fluorescent antibodies to target *Legionella pneumophila* in thin, laboratory grown biofilms. A similar approach has been taken by Szwerinski *et al.* (1985) to identify nitrifying bacteria in thick biofilms. We have used fluorescent antibodies to target *Klebsiella pneumoniae* in mixed species biofilms. Experience has shown that this method is best applied in relatively clean samples and is probably inappropriate for most distribution system samples due to nonspecific binding of the antibody, interference of inorganic matter and difficulty in ascertaining penetration of the biofilm by the antibody.

Molecular Probes in Biofilm Research

Increasingly, environmental microorganisms, including those in biofilms, are detected by so-called 'molecular' methods. A broad interpretation of the term includes methods that do not require bacterial growth on selective or non-selective media. A narrow interpretation is limited to methods that target nucleic acids or proteins. Either whole cells or purified cell extracts of DNA, RNA, or proteins can be analysed. Whole cells can be extracted and/or concentrated from samples or analysed *in situ*. *In situ* detection of undisturbed whole cells, especially in biofilms, is essential in order to determine the spatial distribution of species. A list of molecular tools includes nucleic acid and protein stains, physiological indicators, labelled antibodies, nucleic acid amplification, nucleic acid probes and gel electrophoresis of nucleic acids and proteins. An integrated molecular approach to studying planktonic or biofilm bacteria would include methods for determining their identity, abundance, and physiological status in single or parallel assays.

Molecular probes are gaining popularity in drinking water biofilm studies. Because of current limitations in their practical application there are still few reports of successful use in these samples, and routine use by the industry is still not feasible. Szewzyk et al. (1994) suggested that oligonucleotide probes can penetrate biofilms found in oligotrophic environments, while labelled antibodies may only detect target cells on the biofilm surface. Manz et al. (1993) inserted glass slides into a Robbins device installed in a drinking water distribution system and noted microcolony formation with phase contrast microscopy within 3–8 weeks. The biofilms were hybridized with oligonucleotide probes and subsequently stained with DAPI. The universal probe (EUB338) was detected in about 70% of the attached cells but only about 40% of the planktonic cells, based on DAPI total counts. This disparity was considered evidence of higher rRNA content of attached cells.

FUTURE DIRECTIONS

The drinking water industry is driven by two major groups, the regulators and the customers. Complying with certain regulations can have a negative effect on the ability of water utilities to comply with other regulations and can impact the public's perception of the water. For example, increasing chlorine concentrations to decrease microbial activity in distribution systems can increase disinfection by-products (a regulatory concern) and result in complaints from customers. To ensure safe, palatable water at an economical cost, it will be necessary to obtain fundamental information on the interaction of a variety of water quality and distribution system parameters.

In the area of microbiology, adaptation and implementation of the suite of new tools being developed in the area of molecular microbial ecology will provide valuable insight on the factors that control biofilm formation in distribution systems. Of key importance will be direct targeting of potentially pathogenic bacteria so that relevant measures of the safety of water can be obtained.

Connected with improved microbiological monitoring methods is the need for sound risk assessment and epidemiological studies on the health effects of a variety of waterborne and water associated organisms. When this information is available, sound decisions on the appropriate means for

ensuring microbiologically and chemically safe water at a reasonable cost
can be made.

REFERENCES

Abernathy, C (1998) The Effect of Corrosion Control Treatments and Biofilm
    Disinfection on Unlined Ferrous Iron Pipes. PhD. Disseration. Dept., Civil
    Engineering. Montana State University, Bozeman, MT.
American Water Works Association Research Foundation-DVGW-Forscungstelle
    (1985) *Internal Corrosion of Water Distribution Systems, Research Report.* AWWA,
    Denver, CO.
Birden, HH Jr, Calabrese, EJ and Stoddard, A (1985) Lead dissolution from soldered
    joints. *Journal of American Water Works Association* 77:66–69.
Block, JC (1992) Biofilms in drinking water distribution systems. In: Biofilms—
    Science and Technology (Eds. Bott, TR, Fletcher, M and Capdeville B), NATO
    Advanced Study Institute, Portugal, Kluwer Publishers, the Netherlands.
Bosch, A, Lucena, F, Diez, JM, Gajardo, R, Blasi, M and Jofre, J (1991) Waterborne
    viruses associated with hepatitis outbreaks. *Journal of American Water Works
    Association* 83:80–83.
Brettar, I and Hvfle, MG (1992) Influence of ecosystematic factors on survival of
    Escherichia coli after large-scale release into lake water mesocosms. *Applied and
    Environmental Microbiology* 58:2201–2210.
Brewer, WS and Carmichael, WW (1979) Microbial characterization of granular
    activated carbon filter systems. *Journal of American Water Works Association* 71:738–740.
Burke, V, Robinson, J, Gracey, M, Petersen, D and Partridge, K (1984a) Isolation of
    Aeromonas hydrophila from a metropolitan water supply: Seasonal correlation with
    clinical isolates. *Applied and Environmental Microbiology* 48:361–366.
Burke, V, Robinson, J, Gracey, M, Peterson, D, Meyer, N and Haley V (1984b)
    Isolation of *Aeromonas* spp. from an unchlorinated domestic water supply. *Applied
    and Environmental Microbiology* 48:367–370.
Burman, NP (1965) Taste and odor due to stagnation and local warming in long
    lengths of piping. *Proceedings of the Society for Water Treatment and Examination*
    14:125–131.
Burman, NP (1973) The occurrence and significance of actinomycetes in water
    supply. In: Actinomycetaeles; Characteristics and Practical Importance (Eds.
    Sykes, G and Skinner, FA), Academic Press, London and New York. pp. 219–230.
Camper, AK (1994) Coliform Regrowth and Biofilm Accumulation in Drinking
    Water Systems: A Review. In: Biofouling and Biocorrosion in Industrial Water
    Systems (Eds. Geesey, GG, Lewandowski, Z and Flemming, HC), CRC-Lewis
    Publishers, Boca Raton, FL.
Camper, AK (1996) Factors Limiting Microbial Growth in the Distribution System:
    Pilot and Laboratory Studies. American Water Works Association Research
    Foundation, Denver, CO.
Camper, AK, LeChevallier, MW, Broadaway, SC and McFeters, GA (1985) Growth
    and persistence of pathogens on granular activated carbon filters. *Applied and
    Environmental Microbiology* 50:1378–1382.
Camper, AK, LeChevallier, MW, Broadaway, SC and McFeters, GA (1986) Bacteria
    associated with granular activated carbon particles in drinking water. *Applied and
    Environmental Microbiology* 52:434–438.

Camper, AK, McFeters, GA, Characklis, WG and Jones, WL (1991) Growth kinetics of coliform bacteria under conditions relevant to drinking water distribution systems. *Applied and Environmental Microbiology* **57**:2233–2239.

Camper, AK, Jones, WL and Hayes, JT (1996) Effect of growth conditions and substratum composition on the persistence of coliforms in mixed population biofilms. *Applied and Environmental Microbiology* **62**:4014–4018.

Carson, LA, Bland, LA, Cusick, LB, Favero, MS, Bolan, GA, Reingold, AL and Good, RC (1988) Prevalence of nontuberculous mycobacteria in water supplies of hemodialysis centers. *Applied and Environmental Microbiology* **54**:3122–3125.

Carson, LA, Petersen, NJ, Favero, MS and Aguero, SM (1978) Growth characteristics of atypical mycobacteria in water and their comparative resistance to disinfectants. *Applied and Environmental Microbiology* **36**:839–846.

Centers for Disease Control (1985) Detection of elevated levels of coliform bacteria in a public water supply. *Morbidity and Mortality Weekly Report* **3**:142.

Characklis, WG (1988) Bacterial Regrowth in Distribution Systems. American Water Works Association Research Foundation, Denver, CO.

Chen, CI, Griebe, T, Srinivasan, R and Stewart, PS (1993) Effects of various metal substrata on accumulation of *Pseudomonas aeruginosa* biofilms and the efficacy of monochloramine as a biocide. *Biofouling* **7**:241–251.

Colbourne, JS, Dennis, PJ, Trew, RM, Berry, C and Vesey, G (1988) *Legionella* and public water supplies. In: Proceedings International Conference on Water and Wastewater Microbiology, Newport Beach, CA.

Colbourne, JS, Dennis, PJ, Keevil, W and Mackerness, C (1991) The operational impact of growth of coliforms in London's distribution system. In: Proc. Water Qual. Technol. Conf. American Water Works Association Research Foundation, Denver, CO.

Collins, CH, Grange, JM and Yates, MD (1984) Mycobacteria in water. *Journal of Applied Bacteriology* **57**:193–211.

Craun, GF (1986) Recent statistics of waterborne disease outbreaks in the U.S. (1920–1980). In: Waterborne Diseases in the United States (Ed. Craun, GF), CRC Press, Boca Raton, FL.

Craun, GF, Berger, PS and Calderon, RL (1997) Coliform bacteria and waterborne disease outbreaks. *Journal of American Water Works Association* **89**:96–104.

Dennis, JP, Taylor, JA, Fitzgeorge, RB and Bartlett, CLR (1982) *Legionella pneumophila* in water plumbing systems. *Lancet* **24**:949–951.

Donlan, RM and Pipes, WO (1988) Selected drinking water characteristics and attached microbial population density. *Journal of American Water Works Association* **80**:70–76.

duMoulin, GC and Stottmeir, KD (1986) Waterborne mycobacteria: An increasing threat to health. *American Society for Microbiology News* **52**:525–529.

duMoulin, GC, Stottmeier, KD, Pelletier, PA, Tsang AY and Hedley-Whyte, J (1988) Concentration of *Mycobacterium avium* by hospital hot water systems. *Journal of the American Medical Association* **260**:1599–1601.

Engel, HWB, Berwald, LG and Havelaar, AH (1980) The occurrence of *Mycobacterium kansasii* in tap water. *Tubercle* **61**:21–26.

Fischeder, R, Schulze-Robbecke, R and Weber, A (1991) Occurrence of mycobacteria in drinking water samples. *Zentralblatt für Hygiene und Umweltmedizin* **192**:154–158.

Gambassini, L, Sacco, C, Lanciotti, E, Burrini, D and Griffini, O (1990) Microbial quality of the water in the distribution system of Florence. *Aquatic Microbiology* **39**:258–264.

Gardels, MC and Sorg, TJ (1989) A laboratory study of the leaching of lead from water faucets. *Journal of American Water Works Association* **81**:101–105.

Geldreich, EE (1990) Microbiological quality control in distribution systems. In: Water Quality and Treatment (Ed. Pontius, FW), McGraw-Hill, New York, NY.

Geldreich, EE (1991) Opportunistic organisms and the water supply connection. In: Proc. Water Qual. Technol. Conf. American Water Works Association Research Foundation, Denver, CO.

Geldreich, EE and Rice, EW (1987) Occurrence, significance and detection of *Klebsiella* in water systems. *Journal of American Water Works Association* **79**:74–80.

Glover, N, Holtzman, A, Aronson, T, Froman, S, Berlin, OGW, Dominguez, P, Kunkel, KA, Overturf, G, Stelma, G, Smith, Jr C and Yakrus, M (1994) The isolation and identification of *Mycobacterium avium* complex (MAC) recovered from Los Angeles potable water, a possible source of infection in AIDS patients. *International Journal of Environmental Health Research* **4**:63–72.

Haas, CN, Meyer, MA and Paller, MS (1983) The ecology of acid-fast organisms in water supply, treatment, and distribution systems. *Journal of American Water Works Association* **75**:139–144.

Haudidier, K, Paquin, JL, Francais, T, Hartemann, P, Grapin, G, Colin, F, Jourdain, MJ, Block, JC, Cheron, J, Pascal, O, Levi, Y and Miazga, J (1988) Biofilm growth in drinking water networks: A preliminary industrial pilot plant experiment. *Water Science and Technology* **20**:109–115.

Heumann, DW (1989) Solid lead gooseneck slug dispersion in consumer plumbling systems. In: Proceedings Water Quality Technology Conference, American Water Works Association. Denver, CO.

Havelaar, HH, Versteegh, JFM and During, M (1990) The presence of *Aeromonas* in drinking water supplies in the Netherlands. *Zentralblatt für Hygiene und Unweltmedizin* **190**:236–256.

Hicks, RE, Amann, RI and Stahl, DA (1992) Dual staining of natural bacterioplankton with 4',6-diamidino-2-phenylindole and fluorescent oligonucleotide probes targeting kingdom-level 16S rRNA sequences. *Applied and Environmental Microbiology* **58**:2158–2163.

Hudson, LD, Hankins, JW and Battaglia, M (1983) Coliforms in a water distribution system: A remedial approach. *Journal of American Water Works Association* **75**:564–568.

Joret, JC (1994) Control of biodegradable organic matter during drinking water treatment. In: Proceedings International Seminar on Biodegradable Organic Matter. Montreal, Quebec.

Kaustova, J, Olsovski, Z, Kubin, M, Zatloukal, O, Pelikan, M and Hradil, V (1981) Endemic occurrence of *Mycobacterium kansasii* in water supply systems. *Journal of Hygiene, Microbiology and Immunology* **25**:24–30.

Kepner, RL and Pratt, JR (1994) Use of fluorochromes for direct enumeration of total bacteria in environmental samples: past and present. *Microbiological Reviews* **58**:603–615.

Kerneis, A, Deguin, A and Feinberg, M (1994) Modeling applications of the number of microorganisms according to the residence time of drinking water in a distribution system. In: Proceedings International Seminar on Biodegradable Organic Matter. Montreal, Quebec.

Knocke, WR (1988) Soluble manganese removal on oxide coated filter media. *Journal of American Water Works Association* **80**:65–70.

Knocke, WR, Shorney, HL and Bellamy, JD (1994) Examining the reactions between soluble iron, DOC, and alternative oxidants during conventional treatment. *Journal of American Water Works Association* **86**:117–127.

Kramer, MH, Herwaldt, BL, Craun, GF, Calderon, RL and Juranek, DD (1996) Waterborne disease: 1993 and 1994. *Journal of American Water Works Association* **88**:66–80.

LeChevallier, MW (1990) Coliform regrowth in drinking water: A review. *Journal of American Water Works Association* **82**:74–86.

LeChevallier, MW, Babcock, TM and Lee, RG (1987) Examination and characterization of distribution system biofilms. *Applied and Environmental Microbiology* **53**:2714–2724.

LeChevallier, MW, Cawthon, CD and Lee, RG (1988) Factors promoting survival of bacteria in chlorinated water supplies. *Applied and Environmental Microbiology* **54**:649–654.

LeChevallier, MW, Evans, TM, Seidler, RJ, Daily, OP, Merrel, BR, Rollins, DM and Joseph, SW (1982) *Aeromonas sobria* in chlorinated drinking water supplies. *Microbial Ecology* **8**:325–333.

LeChevallier, MW, Lowry, CD and Lee, RG (1990) Disinfecting biofilms in a model distribution system biofilm. *Journal of American Water Works Association* **82**:87–99.

LeChevallier, MW, Lowry, CD, Lee, RG and Gibbon, DL (1993) Examining the relationship between iron corrosion and the disinfection of biofilm bacteria. *Journal of American Water Works Association* **85**:111–123.

LeChevallier, MW, Norton, CD, Camper, A, Morin, P, Ellis, B, Jones, W, Rompre, A, Prevost, M, Coallier, J, Servais, P, Holt, D, Delanoue, A and Colbourne, J (1998) Microbial Impact of Biological Filtration. American Water Works Association Research Foundation. Denver, CO.

LeChevallier, MW, Schulz, W and Lee, RG (1991) Bacterial nutrients in drinking water. *Applied and Environmental Microbiology* **57**:857–862.

LeChevallier, MW, Welch, NJ and Smith, DB (1996a) Factors Limiting Microbial Growth in the Distribution System: Full Scale Experiments. American Water Works Association Research Foundation. Denver, CO.

LeChevallier, MW, Welch, NJ and Smith, DB (1996b) Full-scale studies of factors related to coliform regrowth in drinking water. *Applied and Environmental Microbiology* **62**:2201–2211.

Lee, SH, O'Connor, S and Banerji, BK (1980) Biologically mediated corrosion and its effects on water quality in distribution systems. *Journal of American Water Works Association* **72**:636–645.

Levin, R and Shock, MR (1991) The use of pipe loop tests for corrosion control diagnostics. In: Proceeding Water Quality Technology Conference, American Water Works Association., Denver, CO.

Lin, YE, Vidic, RD, Stout, JE and Yu, VL (1998) *Legionella* in water distribution systems. *Journal of American Water Works Association* **90**:112–121.

Lippy, EC and Waltrip, SC (1984) Waterborne disease outbreaks 1946–1980: A thirty-five year perspective. *Journal of American Water Works Association* **76**:60–67.

Lowther, ED and Moser, RH (1984) Detecting and eliminating coliform regrowth. In: Proceedings Water Quality Technology Conference, American Water Works Association., Denver, CO.

Manz, W, Szewzyk, U, Ericsson, P, Amann, R, Schleifer, KH and Stenstrom, TA (1993) *In situ* identification of bacteria in drinking water and adjoining biofilms by hybridization with 16S and 23S rRNA-directed fluorescent oligonucleotide probes. *Applied and Environmental Microbiology* **59**:2293–2298.

Martin, RS, Gates, WH, Tobin, RS, Sumarah, R, Wolfe, P and Forestall, P (1982) Factors affecting coliform bacterial growth in distribution systems. *Journal of American Water Works Association* **74**:34–37.

McFeters, GA, Yu, FP, Pyle, BH and Stewart, PS (1995) Physiological assessment of bacteria using fluorochromes, a review. *Journal of Microbiological Methods* **21**:1–13.

Morin, P and Camper, AK (1997) Attachment and fate of carbon fines in simulated drinking water distribution system biofilms. *Water Research* **31**:399–410.

Mosley, JW (1966) Transmission of viral diseases by drinking water. In: Transmission of Viruses by the Water Route, pp. 5–23. Interscience, N.Y.

Neden, DG, Jones, RJ, Smith, JR, Kirmeyer, GJ and Foust, GW (1992) Comparing chlorination and chloramination for controlling bacterial regrowth. *Journal of American Water Works Association* **84**:80–88.

Oliveri, VP, Bakalian, AE, Bossung, KW and Lowther, ED (1985) Recurrent coliforms in water distribution systems and the presence of free residual chlorine. In: Water Chlorination, Chemistry, Environmental Impact, and Health Effects (Eds. Jolley, RL, Bull, RJ, Davis, WP, Katz, S, Roberts, MH and Jacobs VA), Lewis Publishers, Boca Raton, FL.

Olson, BH (1982) Assessment and implications of bacterial regrowth in water distribution systems. EPA-6001/52-82-072, U.S. Environmental Protection Agency, Cincinnati, OH.

Payment, P, Richardson, L, Siematycki, J, Dewar, R, Edwardes, M and Franco, E (1991) A randomized trial to evaluate the risk of gastrointestinal disease due to consumption of drinking water meeting current microbiological standards. *American Journal of Public Health* **81**:703–708.

Payment, P, Franco, E and Siemiatycki, J (1993) Absence of relationship between health effects due to tap water consumption and drinking water quality parameters. *Water Science Technology* **27**:137–143.

Pipes, WO (1990) Microbiological methods and monitoring of drinking water. In: Drinking Water Microbiology (Ed. McFeters, GA), Springer-Verlag, New York, NY.

Pyle, BH, Broadaway, SC and McFeters, GA (1995) A rapid, direct method for enumerating respiring enterohemorraghic *Escherichia coli* O157:H7 in water. *Applied and Environmental Microbiology* **61**:2614–2619.

Reilly, KJ and Kippen, JS (1984) Relationship of bacterial counts with turbidity and free chlorine in two distribution systems. *Journal of American Water Works Association* **75**:309–314.

Rice, EW, Johnson, CH, Wild, DK and Reasoner, DJ (1992) Survival of *Escherichia coli* O157:H7 in drinking water associated with a waterborne disease outbreak of hemorrhagic colitis. *Letters in Applied Microbiology* **15**:38–40.

Ridgway, HF, Means, EG and Olson, BH (1981) Iron bacteria in drinking-water distribution systems: Elemental analysis of *Gallionella* stalks using X-ray energy-dispersive microanalysis. *Applied and Environmental Microbiology* **41**:288–292.

Rippey, SR and Cabelli, VJ (1979) Membrane filter procedure for enumeration of *Aeromonas hydrophila* in fresh waters. *Applied and Environmental Microbiology* **38**:108–113.

Robinson, PJ, Walker, JT, Keevil, CW and Cole, J (1995) Reporter genes and fluorescent probes for studying the colonisation of biofilms in a drinking water supply line by enteric bacteria. *FEMS Microbiology Letters* **129**:183–188.

Rogers, J, Dowsett, AB, Dennis, PJ, Lee, JV and Keevil, CW (1994) Influence of materials on biofilm formation and growth of *Legionella pneumophila* in potable water systems. *Applied and Environmental Microbiology* **60**:1842–1851.

Rogers, J and Keevil, CW (1992) Immunogold and fluorescein immunolabelling of *Legionella pneumophila* within an aquatic biofilm visualized by using episcopic differential interference contrast microscopy. *Applied and Environmental Microbiology* **58**:2326–2330.

Rogers, J and Keevil, CW (1995) Survival of *Cryptosporidium parum* oocysts in biofilm and planktonic samples in a model system. In: Protozoan Parasites and Water (Eds. Betts, WB, Casemore, P, Fricker, C, Smith, H and Watkins), Royal Society of Chemistry, U.K.

Rollinger, Y and Dott, W (1987) Survival of selected bacterial species in sterilized activated carbon filters and biological activated carbon filters. *Applied and Environmental Microbiology* **53**:777–781.

Roszak, DB, Grimes, BJ and Colwell, RR (1983) Viable but nonrecoverable stage of *Salmonella enteritidis* in aquatic systems. *Canadian Journal Microbiology* **30**:334–338.

Schreppel, CK, Fredericksen, DW and Geiss, AA (1997) The positive effects of corrosion control on lead levels and biofilms. In: Proceedings Water Quality Technology Conference, American Water Works Association, Denver, CO.

Schultze-Robbecke, R and Fischeder, R (1989) Mycobacteria in biofilms. *Zentralblatt für Hygiene und Unweltmedizin* **188**:385–390.

Schultze-Robbecke, R, Janning, B and Fischeder, R (1992) Occurrence of myco-bacteria in biofilm samples. *Tubercle and Lung Disease* **73**:141–144.

Servais, P, Billen, G, Ventresque, C and Bablon, GP (1991) Microbial activity in GAC filters at the Choisy-le-Roi treatment plant. *Journal of American Water Works Association* **83**:62–68.

Smith, DB, Hess, AF and Hubbs, SA (1990) Survey of distribution system coliform occurrences in the United States. In: Proceedings Water Quality Technology Conference American Water Works Association Research Foundation, Denver, CO.

Sobsey, MD, Dufour, AP, Gerba, CP, LeChevallier, MW and Payment, P (1993) Using a conceptual framework for assessing risks to health from microbes in drinking water. *Journal of American Water Works Association* **85**:44–48.

Sobsey, MD, Shields, PA, Hauchman, FS, Davis, AL, Rullman, VA and Bosch, A (1988) Survival and persistence of Hepatitis A virus in environmental samples. In: Viral Hepatitis and Liver Disease (Ed. Zuckerman, AJ), Alan R. Liss, Inc., N.Y.

Stout, JE, Yu, VL and Best, MG (1985) Ecology of Legionella pneumophila within water distribution systems. *Applied and Environmental Microbiology* **49**:221–228.

Szewzyk, R and Manz, W (1992) Survival of pathogenic bacteria in biofilms of water bacteria. Abstract of the Sixth International Symposium on Microbial Ecology, Barcelona, Spain. #P2-04-21.

Szewzyk, U, Manz, W, Amann, R, Schleifer, K-H and Stenstrvm, TA (1994) Growth and in situ detection of a pathogenic *Escherichia coli* in biofilms of a heterotrophic water-bacterium by use of 16S- and 23S-rRNA-directed fluorescent oligonucleo-tide probes. *FEMS Microbiology Ecology* **13**:169–176.

Szwerinski, H, Gaiser, S and Barkdte, D (1985) Immunofluorescence for the quantitative determination of nitrifying bacteria: Interference of the test in biofilm reactors. *Applied Microbiology and Biotechnology* **21**:125–128.

Tobin, RS, Smith, DK and Lidsay, JA (1981a) Effects of activated carbon and bacteriostatic filters on microbiological quality of drinking water. *Applied and Environmental Microbiology* **41**:646–651.

Tobin, JOH, Swann, RA and Bartlett, CLR (1981b) Isolation of *Legionella pneumophila* from water systems: methods and preliminary results. *British Medical Journal* **282**:515–517.

van der Kooij, D (1988) Properties of aeromonads and their occurrence and hygienic significance in drinking water. *Zentralblatt für Hygiene und Unweltmedizin* **187**:1–17.

van der Kooij, D (1992) Assimilable organic carbon as an indicator of bacterial regrowth. *Journal American Water Works Association* **84**:57–65.

van der Wende, E and Characklis, WG (1990) Biofilms in potable water systems. In: Drinking Water Microbiology (Ed. McFeters, GA), Springer-Verlag, New York.

Vasconcelos, JJ, Boulos, PF, Grayman, WM, Laurent, K, Wable, O, Biswas, P, Bhari, A, Rossman, LA, Clark, RJ and Goodrich, JA (1996) Characterization and Modeling of Chorine Decay in Distribution Systems. American Water Works Association Research Foundation, Denver, CO.

Versteegh, JFM, Havelaar, AH, Hoskstra, AC and Visser, A (1989) Complexing of copper in drinking water samples to enhance recovery of *Aeromonas* and other bacteria. *Journal Applied Bacteriology* **67**:561–566.

Victoreen, HT (1974) Control of water quality in transmission and distribution mains. *Journal of American Water Works Association* **66**:369–370.

von Reyn, CF, Waddell, RD, Eaton, T, Arbeit, RD, Maslow, JN, Barber, TW, Brindle, RJ, Gilks, CF, Lumio, J, Lahdevirta, J, Ranki, A, Dawson, D and Falkinham III, JO (1993) Isolation of *Mycobacterium avium* complex from water in the United States, Finland, Zaire, and Kenya. *Journal Clinical Microbiology* **31**:3227–3230.

von Reyn, CF, Maslow, JN, Barber, TW, Falkinham III, JO and Arbeit, RD (1994) Persistent colonisation of potable water as a source of *Mycobacterium avium* infection in AIDS. *Lancet* **343**:1137–1141.

Wadowsky, RM and Yee, RB (1983) Satellite growth of *Legionella pneumophila* with an environmental isolate of *Flavobacterium breve*. *Applied and Environmental Microbiology* **46**:1447–1449.

Wadowsky, RM and Yee, RB (1985) Effect of non-legonellaceae bacteria on the multiplication of *Legionella pneumophila* in potable water. *Applied and Environmental Microbiology* **49**:1206–1210.

Wadowsky, RM, Yee, RB, Mezmar, L, Wing, EJ and Dowling, JN (1982) Hot water systems as sources of *Legionella pneumophila* in hospital and non-hospital plumbing fixtures. *Applied and Environmental Microbiology* **43**:1104–1110.

Wagner, M, Amann, R, Lemmer, H and Schleifer, KH (1993) Probing activated sludge with oligonucleotide specific for Proteobacteria: inadequacy of culture-dependent methods for describing microbial community structure. *Applied and Environmental Microbiology* **59**:1520–1525.

Warburton, DW, McCormick, JK and Bowen, B (1994) Survival and recovery of *Aeromonas hydrophila* in water: Development of methodology for testing bottled water in Canada. *Canadian Journal Microbiology* **40**:145–148.

Warnecke, M (1996) MS Thesis. Department of Microbiology, Montana State University, Bozeman, MT.

Wolfe, RL, Lieu, NI, Izaguirre, G and Means, EG (1990) Ammonia oxidizing bacteria in a chloraminated distribution system: Season occurrence, distribution, and disinfection resistance. *Applied and Environmental Microbiology* **56**:451–462.

Yu, FP, Callis, GM, Stewart, PS, Griebe, T and McFeters, GA (1994) Cryosectioning of biofilms for microscopic analysis. *Biofouling* **8**:85–91.

2.2 Detection of Biofouling in Drinking Water Systems

ELIZABETH J. FRICKER

Thames Water Utilities Ltd., Manor Farm Road, Reading RG7 1HD, UK

INTRODUCTION

The fact that biofilms exist in drinking water distribution systems has been well established for over a decade. There are many publications that deal with the problems associated with the presence of these biofilms and the issue has been well described in the previous chapter. The complete characterization of biofilms in terms of the matrix and cellular components has, however, been more difficult to determine, as biofilms are inherently variable. A number of publications have recently emerged that have helped to provide a more thorough and almost entire picture of the biofilm in terms of its development, its constituent parts and its contribution to the ecosystem of the drinking water distribution system. Most of the recent increase in knowledge of biofilm characteristics has resulted from improvements in technology and methodology which have facilitated more in depth investigations. This chapter will describe some of the earlier methods used which are still employed today and the more recent techniques now available, for biofilm detection and investigation.

Lazarova and Manem (1995) wrote an informative review of the methods used for the characterization of biofilms used in waste water and water treatment processes. They describe common and advanced analytical methods currently used for biofilm activity and composition analysis. The biofilms described are employed as integral aspects of the treatment processes. Measurement of the activity of the biofilm is seen as the most important parameter. The authors discussed the merits of a number of microbiology-based techniques including: ATP (adenosine triphosphate) measurement; use of INT [2-(p-iodophenyl)-3-(p-nitrophenyl)-5-phenyl tetrazolium chloride]; light microscopy and scanning electron microscopy (SEM) analysis using computer enhanced technology; confocal scanning laser microscopy (CSLM) examination; INT-dehydrogenase activity (DHA) determination; respiratory activity measurement using CTC (5-cyano-2,

Industrial Biofouling. Edited by J. Walker, S. Surman, J. Jass
©2000 John Wiley & Sons Ltd

3-ditolyl-tetrazolium chloride) and DNA content determination using spectrophotometric, fluorometric or radioisotopic techniques. Whilst the biofilms referred to in this review generally carry out part of a treatment function, nevertheless several of the techniques described have an application in the field of detection of biofilms in water distribution pipes.

An earlier review (Costerton et al. 1987) describes the use of SEM and transmission electron microscopy (TEM) for the study of biofilms on medical biomaterials. In this review, mention is also made of the use of quantitative recovery techniques based on removal of the biofilm from the surface. This approach has also been used for the detection of biofilms on distribution water pipe surfaces.

A project funded by the American Water Works Association Research Foundation (AWWARF; Lu and Huck 1993) reported methods available for the measurement of biomass and biofilm thickness. These methods were principally examined from the point of view of adapting wastewater biofilm process fundamentals to the design and operation of drinking water treatment, but again the techniques employed are relevant to biofilm studies in distribution water pipes. The methods reviewed include: SEM and environmental SEM (ESEM); epifluorescence microscopy using acridine orange staining; ATP determination using colorimetric reactions and spectrophotometric adsorption measurements; INT-based electron transport system activity determination and phospholipid concentration measurement using colorimetric analysis of complexed ammonium molybdate and malachite green.

These authors (Lazarova and Manem 1995; Costerton et al. 1987; Lu and Huck 1993) have provided comprehensive overviews of the methods currently employed in general biofilm analysis. There are perceived advantages and disadvantages to most of the techniques. These, as well as the relevant techniques themselves, mentioned in these reviews and those used in other studies, will be discussed below.

SCANNING ELECTRON MICROSCOPY ANALYSIS

The majority of early studies of biofilm characteristics used SEM as the investigative tool. This technique was used to characterize the biofilms themselves and to study the content and nature of the matrix. A considerable amount of work done in the last two decades has concentrated on measuring the thickness and biomass of biofilms. The thickness may be measured using light microscopy, but SEM provides better resolution and magnification and provides a three-dimensional image (Weber et al. 1978).

Some of the pioneering SEM work was aimed at determining the extent and nature of biofilms on pipe surfaces and suspended particulate matter

(Ridgway and Olson 1981). These studies were focused on the association of bacteria and other microorganisms with particulate matter and especially with the inner surfaces of water mains. Other studies of water distribution systems and the associated bacteria and viruses had already produced an abundance of information on the presence and diversity of these organisms in the planktonic phase (Characklis *et al.* 1986; LeChevallier *et al.* 1987; Olson 1982; Seidler *et al.* 1977; Tuovenin and Hsu 1982). Also measures had been prescribed both at the treatment works and in the distribution system, to control potential pathogenic organisms in potable water, based on the presence of the indicator organism *Escherichia coli*. There were still, however, reports of high numbers of coliform bacteria occasionally appearing in planktonic samples from the distribution system. This phenomenon was thus investigated (Ridgway and Olson 1981). SEM was used to determine the morphology and distribution of microorganisms on the water surfaces and X-ray energy dispersive microanalysis was used to determine the chemical composition of the pipe surface. This work showed that the composition of the attachment surface almost certainly determined to some extent the constituents of the attached microbiological population. The structures observed were described in many cases as being similar to either filamentous or more complex chain-forming microbiological species, found in the environment. Extracellular mucopolysaccharide fibrils (i.e. glycocalyx) were also evident from the SEM micrographs. The work concluded that the occurrence of some of the coliforms in the planktonic samples might have been derived from biofilms on the pipe surface although the detailed nature of the biofilm was not determined. It was noted that cultivation *in vitro* or demonstration of the metabolic activity of the observed microcolonies would confirm their microbial nature.

The preparation of the sample is often seen as the main disadvantage to SEM. The sample is fixed in a gluteraldehyde solution, washed, dehydrated in increasing concentrations of ethanol, dried and mounted on an aluminium stub. This is then treated to produce contrast, usually coated with gold, and examined. There are two alternative drying options, critical point drying or freeze-drying. It has been noted that freeze-drying may be more advantageous than critical point drying (Hayat 1975). A further disadvantage of SEM is the distortion the biofilm undergoes during the dehydration stage. The more recently developed ESEM overcomes this issue by permitting direct observation of wet preparations of biofilms. There are, however, few ESEM instruments available and it has been suggested that refinements are required to improve the resolution of this technique (Steele *et al.* 1994). Overall, SEM or ESEM is generally preferred over conventional light microscopy despite the drawbacks of the sample preparation, because it provides a more detailed and in depth view of the biofilm.

THE 'ROBBINS' DEVICE

In order to obtain information relating to biofilms *in situ*, a popular addition to biofilm studies is the use of a 'Robbins' device. The device facilitates investigation of various materials for their biofilm formation potential and characteristics (Manz *et al.* 1993; Amann *et al.* 1996; Kalmbach *et al.* 1997a, b). The device consists of a cylinder, usually made out of stainless steel with threaded holes in the side. Stainless steel screws with glass or polyethylene slides are inserted through the holes of the side-wall. The assembled device may be installed in the distribution system and the mounted slides may be removed independently at various times. Any biofilm present may then be examined using the appropriate technique.

PREPARATION OF BIOFILMS FOR ANALYSIS

In order to examine the bacterial constituents of a biofilm, it has been customary to scrape the biofilm from the surface, homogenize the sample and examine sub-samples using culture techniques (Costerton *et al.* 1987). The tool used for removal may be a sterile dental probe (Walker and Morales 1997), a sterile cell scraper (LeChevallier *et al.* 1987) or other suitable sterile implement which will detach the biofilm from the surface, in preparation for further processing. Some workers have carried out direct sonication of the material (van den Hoven *et al.* 1994), followed by further sonication to homogenize the sample (Röske *et al.* 1998). A vacuum sampling method for removal of biofilm from PVC drums has also been described (van der Wende *et al.* 1989). The biofilm was vacuumed off using phosphate buffer before being homogenized and analysed using the spread plate technique on R_2A agar. This technique has not been widely utilized, thus its effectiveness is not known.

One of the disadvantages of biofilm removal is the potential for the cells to lyse or become damaged in some way. In addition, it is time consuming and is unable to provide a true representation of the *in situ* location and status of the entire biofilm. However, much of the information about biofilms that is currently available has been gained using removal techniques followed by further analysis.

BIOFILM ANALYSIS USING CULTURE

The study of the microbiology of potable water has traditionally been based on a public health perspective, and the initial approaches for the study of biofilms have been based on the same principle. The culture methods employed for biofilm analysis have been directly transferred from those developed for detection of organisms in planktonic phase samples. Thus the most common bacteriological assay used for analysis of biofilms is the culture of heterotrophic bacteria. Conventional culture analysis of scrape samples has also provided information on the presence of certain species of organisms (Percival *et al.* 1998).

An example of the characterization of biofilms based on culture analysis is described by LeChevallier *et al.* (1987). This comprehensive study examined distribution system biofilms for the presence of coliform bacteria. A number of researchers had previously identified a range of different bacterial species, yeasts and fungi at varying concentrations on different pipe material surfaces, but the occurrence data for coliform bacteria was deficient. Some studies had demonstrated the presence of *Klebsiella* spp. (Seidler *et al.* 1977), *Escherichia coli* (Olson 1982), *Enterobacter cloacae* and *Serratia marcescens* (Victoreen 1977) in various distribution system biofilms but this evidence had not been correlated with the presence of coliforms in potable water samples. Some other researchers failed to isolate coliforms in distribution biofilms (Tuovinen and Hsu 1982; Charaklis *et al.* 1986). Thus this study (LeChevallier *et al.* 1987) endeavoured to determine the best method for recovery and detection of heterotrophic bacteria and the occurrence of coliforms in distribution water pipe biofilms. In order to investigate the general aspects of the biofilms analysed, SEM using critical point drying was employed.

A number of different sample-types including particulates, surface scrapings, floc material, sediments and tubercles were analysed. Particle-associated bacteria were analysed by initially trapping them on filters (12 μm) through which 500 to 2000 litres of water had passed. The particles were removed from the membrane by shaking in buffer and the bacteria were dissociated from the particles by homogenization. Floc samples were homogenized directly and tubercles and sediment crushed and homogenized.

All sample types were analysed for coliforms using the five-tube most-probable-number (MPN) method. The heterotrophic plate count (HPC) analysis was carried out using R_2A agar. Experiments were performed comparing R_2A incubated at room temperature for 7 days, with the standard US method of pour plate incubated at 35° C for 48 h (APHS 1985). The former method recovered significantly higher numbers of bacteria and was, therefore, used for the majority of the study. It is now generally recognized in the USA that the optimum recovery of heterotrophic bacteria is obtained from biofilms using this method. In the UK both yeast extract agar (YEA) and R_2A are used. Overall the study concluded that coliforms were present in the distribution system in tubercles on iron pipes. Heterotrophs were also present in unique populations at different locations.

A residual of 1.0 mg L^{-1} of free chlorine failed to eliminate these bacteria, presumably due to the presence of the protective biofilm.

ATOMIC FORCE MICROSCOPY ANALYSIS

A report by Steele *et al.* (1994) describes a method for the examination of biofilms on stainless steel using atomic force microscopy (AFM). AFM is a surface profile scanning technique that uses a sharp tip mounted on a flexible cantilever to detect surface topography down to the atomic level. The technique produces a three-dimensional image of the biofilm. Previous studies (Gould *et al.* 1990; Kasas *et al.* 1993) used AFM with samples prepared as for SEM. Steele *et al.* (1994) developed the method to enable biofilms to be viewed in a hydrated state. There are relatively few reports of the use of AFM for biofilm imaging, although Steele *et al.* (1994) refer to some work carried out by Bremer *et al.* (1992), examining the topography of a hydrated bacterial biofilms on a copper surface. In general the specialized nature of the technique and instrumentation would prevent this method from being applied routinely for detection of biofilms. Although AFM may be an advance on SEM, the data generated does not provide microbial identification, viability or enumeration.

BIOFILM ACTIVITY MEASUREMENT

Bacterial numbers and biofilm thickness ratios will vary according to the nature of the bacteria present, their physiology and the environment (Characklis 1990). Thus, it is generally accepted that direct measurement of biofilm activity is preferable. This parameter is applicable to most areas in which biofilms are characterized and may be determined using a number of analytical techniques.

Biofilm activity may be assessed by determining adenosine triphosphate (ATP) activity. ATP is involved in bacterial metabolism and disappears rapidly after cell death. The principle of ATP determination is usually based on the measurement of the quantity of light produced when luciferin is oxidized in the presence of ATP and the enzyme luciferase. Although the levels of ATP present remain constant after freezing, the method is not standardized. One specific disadvantage is that the assay is sensitive to the extraction procedure employed and this is particularly relevant for extraction of ATP from biofilm samples. In addition, ATP determination is not specific to bacteria, as it does not differentiate between prokaryotes and eukaryotes.

A more useful method for estimation of biofilm activity, which is applicable in a number of situations, is the determination of electron transport system activity. This may be measured using one of a number of chemical indicators such as the redox dyes triphenyl tetrazolium chloride (TTC), 2-(p-iodophenyl)-3-(p-nitrophenyl)-5-phenyl tetrazolium chloride (INT) or 5-cyano-2, 3-ditolyl tetrazolium chloride (CTC). These substances are colourless, and after reduction under defined conditions, are transformed into red coloured monoformazans. CTC has the added advantage that it emits red fluorescence under ultraviolet light, which permits epifluorescence detection. Assessment using these chemical indicators provides a measure of cumulative respiratory activity and has been linked to determination of viability (Zhang and Bishop 1994). An advantage to measuring the electron transport system activity over ATP determination, is that it does not require disruption of intracellular or extracellular processes. The chemical indicators are reduced by the dehydrogenases of respiring bacteria to an insoluble (fluorescent) formazan product, which is then detected by microscopy. This technique has been employed in environmental sample analysis of planktonic and sessile cultures (Blenkinsopp and Lock 1990; Rodriguez et al. 1992; Schaule et al. 1993; Lazarova et al. 1994; Schwartz et al. 1998a).

TTC measurement has been shown to correlate well with ATP content, oxygen consumption and total cell number (Klapwijk et al. 1974). However, TTC is very sensitive to the presence of oxygen. INT does not suffer from this problem and correlates well with the other parameters. CTC has a very low redox potential and thus abiotic reduction is not a significant problem. TTC, CTC and INT determinations are sensitive to biochemical changes and depend on the physiological state of the organisms. Thus, the analysis is most appropriate for assessment of stable bacterial populations. Some workers have developed the INT-method to improve the microscopic detection. Addition of yeast extract (to stimulate bacterial activity) and nalidixic acid (to inhibit cellular division) to the incubation medium has been shown to improve the threshold of detection (Maki and Remsen 1981).

MICROSCOPY-BASED DETECTION METHODS

Direct detection of bacteria within biofilms may be achieved by utilization of various microbiological stains, which are visible under epifluorescence microscopy. The most commonly used stain is acridine orange (AO) and the technique employed is described as the acridine orange direct count (AODC) method. AO is a nucleic acid stain, which when used for labelling microorganisms is generally recognized as providing a measure of total number of bacterial cells present. It has been suggested that variation in the colour of the cells may discriminate between active (orange–red fluorescence) and non-active (green) cells. However, this is not a reliable measure or discrimination, as the growth conditions, variable staining procedures, fixation techniques and the presence of background noise will influence the colour of the stained cells and make counting imprecise (Turakhia and Characklis 1989). Thus AO should only be used to enumerate total cell numbers (McFeters *et al.* 1991). A 0.02% solution of AO in 2% formaldehyde is generally used and visualization of AO stained cells is achieved using an excitation filter of 450–490 nm. A number of studies have used this stain to examine the characteristic distribution of bacterial cells in biofilms (Kogure *et al.* 1984; McFeters *et al.* 1991; Zhang and Bishop 1994). Also, Rhodamine 123 may be used as an activity stain and total cell numbers quantified after fixation of samples using the nucleic acid stain propidium iodide (PI) (Camper *et al.* 1998). This compound penetrates inactive cell membranes and binds to double-stranded nucleic acid. PI is excited by blue light and will fluoresce red.

A stain more commonly utilized now for detection of total cell numbers is 4′,6-diamidino-2-phenylindole (DAPI). This nucleic acid stain is easy to visualize and non-discriminatory in terms of viability. It interchelates between the double strands of DNA. The staining procedure is simple to perform. The DAPI stock solution should be prepared and stored at 1–2 mg ml^{-1} in methanol at 2–8° C, for up to 1 month. For routine use the working solution is generally prepared on a daily basis at a concentration of 1–2 μg ml^{-1}. The staining is carried out at room temperature for 5–20 min. False positive signals linked to the presence of organic matter, as has been seen with detection using the tetrazolium salts, do not occur. DAPI is excited by short wavelength light and fluoresces blue, has good stability and slow extinction. DAPI may also be useful for studying early biofilm development.

APPLICATION OF DETECTION METHODS IN ENVIRONMENTAL SAMPLES

Kalmbach *et al.* (1997a) investigated the dynamics of biofilm formation in drinking water using microscopy-based techniques. One of the methods used to determine total cell counts and actively respiring cell numbers was DAPI (1 μg ml^{-1}) in combination with CTC (0.5 mM), respectively. They also developed a protocol for determining viable counts directly in the biofilm. They modified the direct count method of Kogure *et al.* (1984) by the addition of pipemidic acid. They initially investigated the performance of several DNA synthesis inhibiting antibiotics; nalidixic acid, pipemidic acid and piromidic acid, using pure cultures. They showed that nalidixic and piromidic acid did not effectively suppress cell division but that pipemidic acid was effective. It has however been noted (Schaule *et al.* 1993), that it may be difficult to determine which cells have undergone elongation, in response to the antibiotic, in an environmental sample. Environmental samples possess cells of many different lengths, which may overlap or aggregate, making result interpretation difficult. As with all techniques, transferral of the proposed method to environmental samples can prove problematic due to the complex nature of the matrices.

Rodriguez *et al.* (1992) demonstrated the use of CTC reduction for native environmental sample investigation and showed the suitability of the method for detection of actively respiring sessile and planktonic bacteria. They showed that the addition of R_2A medium to the CTC incubation medium improved the detection of the bacteria. Schaule *et al.* (1993) showed that CTC reduction was a rapid and sensitive method for quantification and localization of viable planktonic and sessile bacteria in drinking water. The CTC assay proposed by Rodriguez *et al.* (1992) was modified by optimizing at a reduced CTC concentration (0.5 mM incubated in the dark for 1 h at room temperature). Schwartz *et al.* (1998a) showed that freshly prepared CTC (5 mM), with the addition of pyruvic acid (2 mM), incubated for 3 h at room temperature in the dark provided the maximum detection of respiring bacteria in biofilms.

Schaule *et al.* (1993) also demonstrated the staining of biofilms directly on coupons and slides. The microscopic examination and detection was enhanced by the use of digital image analysis which quantifies the fluorescence intensity relative to an arbitrary standard. As with all studies, which have compared CTC with culture for detection of viable cell numbers, the CTC method yielded significantly greater estimates of cell viability than standard R_2A plate counts. Overall, application of the majority of techniques mentioned here, to environmental samples, has shown more variability with the result than that obtained with pure culture investigation, as might be expected.

FOURIER TRANSFORM INFRARED SPECTROSCOPY

An investigation into biochemical methods for biofilm monitoring (Amann *et al.* 1996) described a spectroscopic method known as Fourier transform infrared spectroscopy (FTIR), that has recently been optimized for analysis of biological molecules. Biofilms grown on different materials can be measured without sample preparation and destruction of the biofilm. The method is described in detail (Amann *et al.* 1996), but essentially it is based on excitation and vibration of molecules by absorption of electromagnetic radiation. The associated technique of attenuated total reflection (ATR) using the multiple attenuated reflection format is used to produce a spectrum, that can be related to corresponding molecular structures. It is possible to distinguish between peptides, proteins, polysaccharides, phospholipids and nucleic acids. Thus the development of biofilms, based on changes in molecular structure, can be observed. Changes in cell number may be monitored, for example, by studying the amide II band and the dynamics of the extracellular polysaccharides (EPS) development can also be followed. It is suggested that the control of biofilm development or the effects of disinfection may be monitored using FTIR. In addition, the availability of reference spectra for a number of taxonomic species and groups, enables the specific characterization of biofilms to be determined.

GAS CHROMATOGRAPHY–MASS SPECTROMETRY

A method that has been employed in the detection of some bacteria in biofilms, is gas chromatography–mass spectrometry (GC–MS). The technique is based on measurement of specific fatty acid biomarkers. In a study (Walker *et al.* 1993) investigating the colonization of plumbing materials by *Legionella pneumophila*, GC–MS was used. *L. pneumophila* has a complex fatty acid composition, which includes some fatty acids that are unique to the species. This fact facilitates selective detection of the bacterium in a mixed population. GC–MS was also utilized in some work that investigated the relationship between corrosion of surfaces in contact with biofilm and the production of EPS (Beech and Gaylarde 1991). The technique was used to analyse for the presence of specific sugars in order to characterize the EPS.

IN SITU HYBRIDIZATION

One of the advantages of examining biofilms using the tetrazolium salts is the *in situ* nature of the analysis. Another technique, which enables direct detection and characterization of biofilms, is the molecular-based method of *in situ* hybridization. This technique uses fluorescently labelled, nucleic acid targeted oligonucleotide probes. Probes may be directed against the DNA, or more commonly, the rRNA. The selection of the target region and probe determines the specificity of the analysis. Thus for example, species or genus specific probes may be selected. Labelling of different probes with various fluorescent markers permits detection of several targets simultaneously. The sensitivity of the method suggests that at least 10^5 ribosomes per cell need to be present in order for the probes to produce a positive result. However, several methods for improving probe sensitivity are becoming available. A study by Manz *et al.* (1993) showed that a range of different probes directed against domains and sub-classes may be used to detect different organisms. They used the *in situ* technique to provide qualitative information on the physiological state of the target organisms. As with many studies, the 'Robbins' device was used to provide biofilm samples. Detection using 16S and 23S rRNA-directed probes showed that higher numbers of the microorganisms associated with biofilms were bound by the probes, than those cells in the planktonic phase. They were able to demonstrate that microcolonies with fewer than 50 cells consisted of mixed populations of bacteria.

The *in situ* technique is relatively simple to perform. The skill lies in the design and selection of the appropriate probe sequence and optimization of the reaction conditions. There are many published sequences available which select for the various domains, subclasses, indicator bacteria and pathogens (Manz *et al.* 1993; Amann *et al.* 1996; Kalmbach *et al.* 1997a, b). This area of biofilm characterization has received considerable attention in recent years. A number of studies have been published which have examined the population characteristics of biofilms (Schwartz *et al.* 1998a), isolated new bacterial species (Kalmbach *et al.* 1997b), determined colonisation patterns (Robinson *et al.* 1995) and performed specific detection (Manz *et al.* 1993; Manz *et al.* 1995; Schwartz *et al.* 1998b).

An extensive review of fluorochromes and their use in the physiological assessment of bacteria, is given by McFeters *et al.* (1995). Some of the fluorochromes most commonly employed in fluorescence detection with oligonucleotide probes are tetramethylrhodamine-5-isothiocyanate (TRITC), fluoroscein isothiocyanate (FITC), 5(6)-carboxy-fluorescein-*N*-hydroxysuccinimide ester (FLUOS) and the indo-carbocyanine dyes, Cy3 or Cy5.

OTHER MOLECULAR TECHNIQUES

Another molecular technique which has been employed in biofilm studies, is the polymerase chain reaction (PCR). As with *in situ* hybridization, the target sequences may be selected to facilitate detection at the genus or species level. The PCR cannot be used to indicate the viability of the detected cell, as it will amplify the DNA from both dead and living cells. It is not uncommon to combine the PCR with *in situ* hybridization for detection of specific organisms (Schwartz *et al.* 1998a). In this study, following the PCR, southern blot hybridization was used prior to confirmation by *in situ* hybridization. In this case, the oligonucleotide probes used for the *in situ* hybridization step were labelled with digoxigenin. Detection was thus performed using an alkaline phosphatase labelled digoxigenin-specific antibody and a chemiluminescent substrate.

Robinson *et al.* (1995) developed some intricate methods for following the attachment and development of *Escherichia coli* in biofilms. The initial section of the study involved the transformation of *E. coli* with a plasmid, containing anaerobically induced *nirB* promoter, fused to the *lacZ* reporter gene. The transformed *E. coli* was then grown in a chemostat and detected using fluorescent probes, targeting β-galactosidase and β-glucuronidase. Observation of the total biofilm flora was enhanced by the use of differential interference contrast (DIC) microscopy. The images were captured using a black and white charged-couple device camera and then further enhanced using an image analysis transputer card and software.

Another development that has been applied to the *in situ* hybridization technique is confocal laser scanning microscopy. This method of detection requires a specific epifluorescence microscope designed to take electronic images of the fluorescently labelled targets. These images are then processed using the computer software, which usually accompanies the instrument. A 3-D image providing a picture of the spatial distribution of the bacteria within the biofilm is obtained (Röske *et al.* 1998).

SUMMARY OF TECHNIQUES

Investigation of biofilms and surfaces using SEM has provided a significant quantity of data relating to the nature of the biofilm and contact surfaces. It has proved beneficial in providing a detailed in depth view of the biofilm. It is still an applicable technique in some studies today, usually in combination with further identification techniques.

Removal of the biofilm from the surface under investigation has facilitated culture studies of the biofilm population and has provided evidence of the constituent microorganisms. Heterotrophic plate count analysis remains part of many studies investigating, for example, the effect of disinfectants or biocides on biofilms. Any culture method used for recovery of bacteria, from either biofilms or planktonic water samples, will selectively detect those bacteria that are able to grow under the given conditions (Reynolds and Fricker 1999). The conditions may be varied in order to attempt to obtain the largest range of culturable cells. The medium and temperature selected should depend partly on the water source and partly on the target organisms. The time of incubation and the number of stressed bacteria present will also affect overall recovery. Care must be taken when applying methods developed with pure cultures to environmental samples, due to the inherent variability of the latter.

The study of biofilm activity is best achieved using CTC as this facilitates epifluorescence microscopy analysis. It is most usefully combined with total cell count measurement using, for example, DAPI.

More advanced techniques such as *in situ* hybridization also have their applications. It is important to continually expand and develop techniques in order to provide the most comprehensive information. Detection of a specific organism *in situ* is now possible, especially with the use of confocal scanning laser microscopy. Methods that detect bacteria, using alternative recovery techniques, will inevitably produce different quantitative data. The selectivity, specificity and sensitivity of the method(s) utilized must always be considered when comparisons are made between different techniques. There is a large historic database of potable water analysis that provides quantitative information and which has been obtained using culture methods. These data have been utilized as the basis of water treatment effectiveness in relation to public health for nearly a century. The other methods described here, for planktonic and sessile organism analysis, produce further valuable information. There is a need to develop, rapid, sensitive methods for detection of bacteria in potable water samples and many of the techniques are applicable to detection of bacteria in biofilms. When specific genera are detected using probes, then culture conditions can be adapted which will facilitate their growth. It is likely in the future that studies will combine the use of rRNA-targeted probes with some form of culture.

REFERENCES

American Public Health Service (1985) Standard Methods for the examination of water and wastewater, 16th ed. American Public Health Association, Washington, D.C.

Amman, RI, Flemming, HC and Obst, UHG (1996) Biochemical methods for rapid and *in situ* monitoring of biofilms and biofilm-associated micro-organisms during drinking-water distribution. *Water Supply* 14:453–471.

Beech, IB and Gaylarde, CC (1991) Microbial polysaccharides and corrosion. *International Biodeterioration* 27: 95–107.

Blenkinsopp, SA and Lock, MA (1990) The measurement of electron transport system activity in river biofilms. *Water Research* 24:441–445.

Bremer, PJ, Geesey, GG and Drake, B (1992) Atomic force microscopy examination of the topography of a hydrated bacterial biofilm on a copper surface. *Current Microbiology* 24:223–230.

Camper, AK, Warnecke, M, Jones, WL and McFeters, GA (1998) Pathogens in Model Distribution System Biofilms. American Water Works Association Research Foundation, Denver.

Characklis, WG (1990) Laboratory biofilm reactors. In: Biofilms (Eds. Characklis, WG and Marshall, KC), Wiley, New York, pp. 55–89.

Characklis, WG, Goodman, D, Hunt, WA and McFeters, GA (1986) Bacterial regrowth in distribution systems. American Water Works Association Research Foundation, Denver.

Costerton, JW, Cheng, K-J, Geesey, GG, Ladd, TI, Nickel, JC, Dasgupta, M and Marrie, TJ (1987) Bacterial biofilms in nature and disease. *Annual Review of Microbiology* **41**:435–464.

Gould, SAC, Drake, B, Prater, CB, Weisenhorn, AL, Manne, S, Hansma, HG and Hansma PK (1990) From atoms to integrated circuit chips, blood cells and bacteria with AFM. *Journal of Vacuum Technology* **1**:369–373.

Hayat, MA (1975) Principles and techniques of scanning electron microscopy. Van Nostrand Reinhold Company (Cited in Lu and Huck, 1993).

Kalmbach, S, Manz, W and Szewzyk, U (1997a) Dynamics of biofilm formation in drinking water: phylogenetic affiliation and metabolic potential of single cells assessed by formazan reduction and in situ hybridisation. *FEMS Microbiology Ecology* **22**:265–279.

Kalmbach, S, Manz, W and Szewzyk, U (1997b) Isolation of new bacterial species from drinking water biofilms and proof of their in situ dominance with highly specific 16S rRNA probes. *Applied and Environmental Microbiology* **63**:4164–4170.

Kasas, S, Gotzos, V and Celio, MR (1993) Observation of living cells using the AFM. *Biophysical Journal* **64**:539–544.

Klapwijk, A, Drent, J and Steenvoorden, JHAM (1974) A modified procedure for the TTC-dehydrogenase test in activated sludge. *Water Research* **8**:121–125.

Kogure, K, Simidu, U and Taga, N (1984) An improved direct viable count method for aquatic bacteria. *Archiv fur Hydrobiologie* **102**:117–122.

Lazarova, V, Pierzo, V, Fontveille, D and Manem, J (1994) Integrated approach for biofilm characterisation and biomass activity control. *Water Science and Technology* **29**:345–354.

Lazarova, V and Manem, J (1995) Biofilm characterisation and activity analysis in water and wastewater treatment. *Water Research* **29**:2227–2245.

LeChevallier, MW, Babcock,TM and Lee, RG (1987) Examination and characterisation of distribution system biofilms. *Applied and Environmental Microbiology* **53**:2714–2724.

Lu, P and Huck, PM (1993) Evaluation of methods for measuring biomass and biofilm thickness in biological drinking water treatment. Proceedings of the Water Quality Technology Conference, Miami, Florida. Part 2 1415–1456. American Water Works Association, Denver.

McFeters, GA, Singh, A, Byun, S, Callis, PR and Williams, S (1991) Acridine orange staining reaction as an index of physiological activity in *Escherichia coli*. *Journal of Microbiological Methods* **13**:87–97.

McFeters, GA, Yu, FP, Pyle, BH and Stewart, PS (1995) Physiological assessment of bacteria using fluorochromes. *Journal of Microbiological Methods* **21**:1–13.

Maki, JS and Remsen, CC (1981) Comparison of two direct-count methods for determining metabolising bacteria in freshwater. *Applied and Environmental Microbiology* **41**:1132–1138.

Manz, W, Amann, R, Szewzyk, R, Szewzyk, U, Stenström, T-A, Hutzler, P and Schleifer, K-H (1995) In situ identification of *Legionellaceae* using 16S rRNA-targeted oligonucleotide probes and confocal laser scanning microscopy. *Microbiology* **141**:29–39.

Manz, W, Szewzyk, U, Ericsson, P, Amann, R, Schleifer, K-H and Stenstrom, T-A (1993) In situ identification of bacteria in drinking water and adjoining biofilms by hybridisation with 16S and 23S rRNA-directed fluorescent oligonucleotide probes. *Applied and Environmental Microbiology* **59**:2293–2298.

Olson, BH (1982) Assessment and implications of material regrowth in water distribution systems. EPA-600/S2-82-072. U.S. Environmental Protection Agency, Cincinnati, Ohio.

Percival, SL, Knapp, JS, Edyvean, R and Wales, DS (1998) Biofilm Development on Stainless Steel in Mains Water. *Water Research* **32**:243–253.

Reynolds, DT and Fricker, CR (1999) Application of laser scanning for the rapid and automated detection of bacteria in water samples. *Journal of Applied Microbiology* **86**:785–795.

Ridgway, HF and Olson, BH (1981) Scanning electron microscope evidence for bacterial colonisation of a drinking-water distribution system. *Applied and Environmental Microbiology* **41**:274–287.

Robinson, PJ, Walker, JT, Keevil, CW and Cole, J (1995) Reporter genes and fluorescent probes for studying the colonisation of biofilms in a drinking water supply line by enteric bacteria. *FEMS Microbiology Letters* **129**:183–188.

Rodriguez, GG, Phipps, D, Ishiguro, K and Ridgway, HF (1992) Use of a fluorescent redox probe for direct visualisation of actively respiring bacteria. *Applied and Environmental Microbiology* **53**:1801–1808.

Röske, I, Röske, K and Uhlmann, D (1998) Gradients in the taxonomic composition of different microbial systems:comparison between biofilms for advanced waste treatment and lake sediments. *Water Science and Technology* **37**:159–166.

Schaule, G, Flemming, H-C and Ridgway, HF (1993) Use of 5-cyano-2,3-ditolyl tetrazolium chloride for quantifying planktonic and sessile respiring bacteria in drinking water. *Applied and Environmental Microbiology* **59**:3850–3857.

Schwartz, T, Hoffmann, S and Obst, U (1998a) Formation and bacterial composition of young, natural biofilms obtained from public bank-filtered drinking water systems. *Water Research* **32**:2787–2797.

Schwartz, T, Kalmbach, S, Hoffman, S, Szewzyk, U and Obst, U (1998b) PCR-based detection of mycobacteria in biofilms from drinking water distribution system. *Journal of Microbiological Methods* **34**:113–123.

Seidler, RJ, Morrow, JE and Bagley, ST (1977) *Klebsielleae* in drinking water emanating from redwood tanks. *Applied and Environmental Microbiology* **33**: 893–900.

Steele, A, Goddard, DT and Beech, IB (1994) An atomic force microscopy study of the biodeterioration of stainless steel in the presence of bacterial biofilms. *International Biodeterioration and Biodegradation* **34**:35–46.

Tuovenin, OH and Hsu, JC (1982) Aerobic and anaerobic micro-organisms in tubercles of the Columbus, Ohio, water distribution system. *Applied and Environmental Microbiology* **44**:761–764.

Turakhia, MH and Characklis, WG (1989) Activity of *Pseudomonas aeruginosa* in biofilms: effect of calcium. *Biotechnology and Bioengineering* **33**:406–414.

van den Hoven, T, van der Kooij, D, Vreeburg, J and Brink, H (1994) Methods to analyse and to cure water quality problems in distribution systems. *Water Supply* **12**:151–159.

van der Wende, E, Characklis, WG and Smith, DB (1989) Biofilms and bacterial drinking water quality. *Water Research* **23**:1313–1322.

Victoreen, HT (1977) Water quality deterioration in pipelines. Proceedings of the Water Quality Technical Conference, Kansas City, Missouri. American Water Works Association, Denver.

Walker, JT and Morales, M (1997) Evaluation of chlorine dioxide (CO_2) for the control of biofilms. *Water Science and Technology* **35**:319–323.

Walker, JT, Sonesson, A, Keevil, CW and White, DC (1993) Detection of *Legionella pneumophila* in biofilms containing a complex microbial consortium by gas chromatography-mass spectrometry analysis of genus-specific hydroxy fatty acids. *FEMS Microbiology Letters* **113**:139–144.

Weber, WJ, Pirbazari, M and Meison, G (1978) Scanning electron microscopy. *Environmental Science and Technology* **12**:817–819.

Zhang, TC and Bishop L (1994) Structure, activity and composition of biofilms. *Water Science and Technology* **29**:926–935.

2.3 Control of Biofouling in Drinking Water Systems

JAMES T. WALKER[1] and STEVEN L. PERCIVAL[2]
[1]*Research Microbiologist, Pathogen Microbiology Group,*
CAMR, Porton Down, Salisbury SP4 0JG, UK
[2]*Senior Lecturer in Medical Microbiology,*
Department of Biological Sciences, The University of
Central Lancashire, Preston, PR1 2HE, UK

INTRODUCTION TO DISINFECTION

Disinfection is used in potable water treatment processes to reduce pathogens to acceptable levels and to prevent public health concerns. However, evidence is mounting to suggest that exposure to chemical by-products formed during the disinfection process may be associated with adverse health effects. Reducing the amount of disinfectant or altering the disinfection process may decrease the exposure to by-product formation, however, these practices may increase the potential for microbial contamination. Therefore, at present, it is necessary for research in the areas of potable water and disinfection to balance the health risks caused by exposure to microbial pathogens with the risks caused by exposure to disinfection by-products.

For biocides to be effective they must possess a number of ideal properties in order for them to be classified as desirable and cost effective. Disinfectants must: (i) control the presence of pathogens in potable water within a certain time period at specified temperatures. This is particularly important as temperature and biocidal activity are loosely related, with biocidal properties reduced at lower temperatures due to loss of enzyme activity; (ii) overcome fluctuations in composition, concentration and conditions of water and wastewater which are to be treated; (iii) non-toxic to humans or domestic animals, nor unpalatable or otherwise objectionable at the required concentrations; (iv) be cost effective and safe and easy to store, transport, handle and apply; (v) have their concentration in the treated water easily and quickly determined; and (vi) persist at a concentration to provide reasonable residual protection against possible recontamination.

Industrial Biofouling. Edited by J. Walker, S. Surman, J. Jass

Often disinfection is the controlling factor in preventing human exposure to disease-causing pathogenic microbes, including viruses, bacteria and protozoa parasites. One ideally suited disinfectant that is used in potable water is chlorine. The reasons for this are due to its availability and cost combined with its ease of handling and measurement, together with historical implications (Geldreich 1996). The process of water chlorination of potable water, initiated at the beginning of this century, provides a safeguard against pathogenic microbes. However, in recent years the finding that chlorination can lead to the formation of by-products that can be toxic or genotoxic to human and animals has lead to a quest for alternative safer disinfectants. Water borne pathogens are in general controlled at standard concentrations of disinfectant. However one area that is of concern is where the water borne planktonic cells have become attached to surfaces to form a biofilm. Such biofilms have been shown to harbour and protect pathogens, lead to fouling of equipment, heat loss and corrosion of the substrata. This is important particularly as the concentration of disinfectant needed to kill pathogens within a biofilm are much higher when compared to planktonic cells. Hence new disinfectants are required that could be both effective in potable water and at the same time control biofilms. Alternatives to chlorine include ozone, chlorine dioxide and chloramines; other options include iodine, bromine, permanganate, hydrogen peroxide, ferrate, silver, UV light, ionizing radiation, high pH and the use of high temperature.

Disinfectant effectiveness is governed by the concentration of the disinfectant (C), which is measured in milligrams per litre, per contact time (T), which is determined in minutes. These CT values for all disinfectants are affected by a number of parameters including temperature, pH, disinfectant demand, cell aggregations, disinfectant mixing rates and organics.

However with disinfection use comes the formation of disinfectant-resistant microbes. In both potable water and waste water it is generally found that the organisms present can be classified under their resistance to disinfection. This is generally: coliforms < virus < protozoan cysts. With regards to coliform inactivation the following order of efficiency with respect to the main disinfectants used in water treatment is seen (Geldreich 1996): ozone > chlorine dioxide >hypochlorous acid > hypochlorite ion >chloramines. Laboratory studies have shown that 3 to 100 times more chlorine is required to inactivate enteric viruses than is needed to kill coliform bacteria when external conditions such as temperature and pH are kept constant.

EFFECTS OF DISINFECTANTS ON BIOFILMS

The choice of treatment for biofilms in potable water is related to whether its prevention or control of accumulation is desirable. Prevention requires sterilization of the incoming water, continuous flow of biocide at high concentration and/or treatment of the substratum that completely inhibits microbial adsorption. The extent to which any treatment can be applied depends on environmental, process and economic considerations.

Generally, when considering biocides for the control of biofilm, a large number of factors have to be looked at. Commonly the rate of cell adsorption to a substratum is directly proportional to the concentration of cells in the bulk water. Therefore, by reducing the cell concentration in the bulk water there will be a decrease in transport rate of cells. Ultimately, the reduced rate of cellular transport will reduce the rate of biofouling. Filtering to remove bacteria to reduce cell numbers can be a very expensive solution, particularly when large volumes of water are used (Percival *et al.* 1997). Disinfection of the incoming drinking water can be effective in minimizing biofilm accumulation. Nevertheless the accumulation of an established biofilm (after a chlorine treatment) is due primarily to growth as well as the transportion and attachment of cells to the already attached cells.

Research investigating the efficiency of disinfectants on biofilms has been performed in laboratory-based studies. From these studies it has been found that microbial attachment to a surface results in decreased disinfection, particularly by chlorine (Ridgway and Olson 1982; LeChevallier *et al.* 1984; Berman *et al.* 1988). It has also been shown that there is a decreased sensitivity to biocides when organisms are attached to a surface with this effect greatly enhanced in older biofilms (LeChevallier *et al.* 1988a, b).

CHLORINE

General Characteristics

Chlorine is the most commonly used biocide for controlling faecal coliforms, coliforms, heterotrophic bacteria and also biofouling within potable water systems. It is usually introduced into water as chlorine gas. Once introduced into water it hydrolyses to (Bitton 1994):

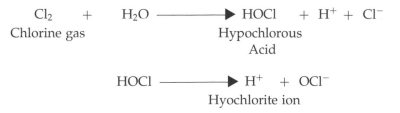

$$Cl_2 \;+\; H_2O \longrightarrow HOCl \;+\; H^+ + Cl^-$$
Chlorine gas   Hypochlorous
                                        Acid

$$HOCl \longrightarrow H^+ \;+\; OCl^-$$
                        Hyochlorite ion

The proportion of HOCl and OCl$^-$ are affected by the pH of water. Free chlorine consists of HOCl or OCl$^-$. Chlorine has been used for many years in potable water because it is an effective method for controlling water quality and biofilms. Concern however, over the toxicity of chlorine and its reaction products has spurred the search for alternative disinfectants as a primary biocide in potable water. The formation of by-products as a result of chlorination of potable water is now recognized as an important hazard particularly due to the risks of colon and bladder cancer. These by-products are referred to as trihalomethanes and are now regarded as carcinogens.

The reaction (depletion) of chlorine in the bulk water is generally referred to as the chlorine demand of water and occurs over time (Characklis and Marshall 1990). The chlorine demand is due to soluble oxidizable inorganic compounds, soluble organic compounds, microbial cells, substratum and particulate in the bulk water. It is now documented that some materials and biofilms found in potable water have chlorine demands which ultimately affect the efficiency of chlorination as a disinfectant.

Mode of Action

Chlorine is known to have two types of effects on bacteria (Bitton 1994). These are:

1. Disruption of cell permeability—chlorine disrupts the integrity of the bacterial cell membrane leading to loss of cell permeability and therefore leaking of proteins, DNA and RNA.
2. Damage to nucleic acids and enzymes.

Effectiveness on Biofilms

The effectiveness of chlorine, within potable water, depends on its ability to inactivate sessile organisms and/or detach significant portions of the biofilm. Chlorine is seen as an effective microbial fouling control biocide because it has been shown to disrupt and loosen biofilms within potable water. It has been found that when chlorine makes contact with a biofilm a number of processes are known to occur. These include: (i) detachment of the biofilm; (ii) dissolution of biofilm components; (iii) disinfection (Characklis 1981).

However, there are a number of factors which are known to influence the rate and extent of the chlorine–biofilm reaction (Characklis 1981):

1. **Turbulent intensity**. Transport of bulk water chlorine to the water–biofilm interface is the first step in the chlorine–biofilm interaction. The transport rate increases with increasing bulk water concentration and turbulence.
2. **Chlorine concentration at the water–biofilm interface**. The transport of chlorine within the biofilm or deposit is a direct function of the chlorine concentration at the interface. Diffusion into the biofilm can

be increased by increasing the chlorine concentration at the bulk water–biofilm interface. High chlorine concentrations for short durations are more effective than low concentrations for long periods assuming the same long term chlorine application rates for both.

3. **Composition of the fouling biofilm.** The reaction of chlorine within the biofilm is dependent on the organic and inorganic composition of the biofilm as well as its thickness or mass. Disinfection in potable water systems is effective at low chlorine concentrations. However, in well developed biofilms, extracellular polysaccharides may compete effectively for available chlorine within the biofilm thereby reducing the chlorine available for killing cells. The substratum may also consume chlorine reducing the concentration attacking the microbial cells.

4. **Fluid shear stress at the water–biofilm interface**. Detachment and sloughing of biofilm, primarily due to fluid shear stress, accompanies the reaction of biofilm with chlorine. Detachment of biofilm due to chlorine treatment has been observed and the rate and extent of removal depend on the chlorine application and on the shear stress at the bulk liquid interface.

5. **pH**. The hypochlorous acid–hypochlorite ion equilibrium may be critical to performance effectiveness. OCl^- apparently favours detachment while $HOCl$ enhances disinfection.

Chlorine is a useful biofouling control compound but, in heavily contaminated waters, is consumed in side reactions (chlorine demand reactions) and is rendered ineffective. Even copper–nickel alloys possess a significant chlorine demand. Therefore, water quality and the substratum composition are the factors that must be considered in choosing a treatment programme to minimize biofilm formation.

The rate at which chlorine is transported through the water phase to the biofilm depends on the concentration of chlorine in the bulk water and the intensity of the turbulence. The chlorine concentration in the bulk water is the net result of the chlorine addition minus the chlorine demand rate of the water. The chlorine concentration at the biofilm–water interface drives the reactions of chlorine within the biofilm. If the chlorine reacts with the biofilm rapidly, the concentration at the interface will be low, and transport of chlorine to the interface may limit the rate of the overall process within the biofilm. By increasing the intensity of turbulence through increased flow rate, both the diffusion in the bulk water and the concentration at the biofilm–water interface will increase.

The transport of chlorine within the biofilm occurs primarily by molecular diffusion. Since the composition of the biofilm is some 96–99% water the diffusivity of chlorine in biofilm is probably a large fraction of its diffusivity in water. In biofilms of higher density or in those containing microbial matter associated with inorganic scales, tubercles or sediment deposits, diffusion of chlorine may be relatively low. Diffusion and the reaction of chlorine in a biofilm determines its penetration and hence its overall effectiveness.

Chlorine reacts with various organic and reduced inorganic components within the biofilm. It can disrupt cellular material (detachment) and inactivate cells (disinfection). In a mature, thick biofilm, significant amounts of chlorine may react with extracellular polysaccharides (EPS), which are responsible for the physiological integrity of the biofilm. With regards to pH, chlorine has been found to be most effective at a pH of 6–6.5, a range at which hypochlorous acid predominates.

Much of the research that has been performed looking at the efficacy of disinfectants against biofilms has generally been done in laboratory-based studies. From a number of studies it has been established that attachment of organisms to surfaces results in a decrease in disinfection by chlorine (Ridgway and Olson 1982; LeChevallier et al. 1984; Berman et al. 1988).

It is accepted that chlorine is to some extent effective against bacteria in the planktonic phase but is less effective against biofilm. Researchers (Knox-Holmes 1993) have shown that low concentrations of chlorine (20 μg l^{-1}) used synergistically with low concentrations of copper (5 μg l^{-1}) prevented growth of micro and macrofouling organisms. Similar effects have been shown with 1 mg l^{-1} of copper and 10 mg l^{-1} of sodium chlorite exposed to *Klebsiella pneumoniae* biofilms for 24 h at 4° C (LeChevallier et al. 1988b).

CHLORAMINES

Chloramines have been proposed as the next best alternative to chlorines due to the public health implications associated with the production of tri-halo-methanes (THM). However, chloramines are not known to be very efficient biocides. In traditional chloramination processess ammonia is added to water first followed by the addition of chlorine in the form of chlorine gas. The conversion rate of free chlorine to chloramines is, as with chlorination, dependant upon pH, temperature, and the ratio of chlorine to ammonia present.

The use of chloramines have been shown to provide a long lasting, measurable disinfectant in potable water. Despite this, research has shown that monochloramines are definitely less effective disinfectants than free chlorine, when compared at comparable low-dose concentrations and short contact periods.

In potable water HOCl reacts with ammonia resulting in the formation of inorganic chloramines:

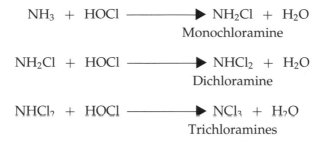

$$NH_3 + HOCl \longrightarrow NH_2Cl + H_2O$$
Monochloramine

$$NH_2Cl + HOCl \longrightarrow NHCl_2 + H_2O$$
Dichloramine

$$NHCl_2 + HOCl \longrightarrow NCl_3 + H_2O$$
Trichloramines

The proportion of these three forms of chloramines depend on the pH of the water with monochloramine predominating at pH >8.5. Monochloramines coexist between pH 4.5 and 8.5 and trichloramine at pH <4.5.

However, a major drawback of using chloramines in potable water and for the control of biofilms is that it results in the formation of low concentrations of nitrites (O'Neill *et al.* 1997). This may result in failures of potable water for nitrite standards, more so in the UK than the US where standards for nitrite levels are less stringent, which are well known to have important implications on human health.

Mode of Action of Chloramines

The mechanism of action of monochloramine may account for its more effective penetration of bacterial biofilms than chlorine (LeChevallier *et al.* 1988a). Monochloramine has been suggested to react rather specifically with nucleic acids, tryptophan and sulphur containing amino acids, but not with sugars such as ribose (Jacangelo and Olivieri 1985).

Effectiveness of Chloramines on Biofilms

Chloramines have been shown to be very effective in suppressing biofilm development, particularly at water temperatures above 15° C. They have been shown to be more effective than chlorine in reducing both sessile coliforms and also heterotrophic bacteria in potable water (Neden *et al.* 1992; Berman *et al.* 1988). In one study it has been found that monochloramines are less effective than free chlorine against planktonic cells (LeChevallier *et al.* 1988b). The reverse was found when these disinfectants were exposed to sessile bacteria.

CHLORINE DIOXIDE

Chlorine dioxide is a strong oxidant formed by a combination of chlorine and sodium chlorine and effectively inactivates bacteria and viruses over a broad pH range (Tanner 1989). Until recently it was used primarily in the textile and pulp/paper industry as a speciality bleach and dye stripping agent. It is often used as a primary disinfectant, inactivating bacteria and cysts. However, it is unable to maintain a residual effect long enough to be useful as a distribution system disinfectant. Despite this it does have advantages over chlorine in that it does not react with precursors to form THMs.

Chlorine dioxide is often commercially sold as stabilized chlorine dioxide that is actually sodium chlorite in a neutral solution. Sodium chlorite is much slower acting and less effective than chlorine and reacts with water to form two by-products. These are chlorite and to some extent chlorate. These compounds have been associated with the oxidation of haemoglobin (Bull 1982) and therefore usage within potable water is restricted to a dosage of 0.5 mg l^{-1}, which is not considered in many cases to be sufficient to provide good disinfection. Other problems associated with chlorine dioxide are the development of taste and odours in some communities. However, chlorine dioxide can oxidize organic compounds such as iron and manganese and suppress a variety of taste and odour problems (Montgomery 1985). Its effectiveness on a number of bacteria, including *Escherichia coli* and *Salmonella* has been noted and found to be equal and greater than free chlorine (Malpas 1973).

Since chorine dioxide is an explosive gas at concentrations above 10% in air it is produced on site by mixing sodium chlorite with either inorganic (e.g. hydrochloric, phosphoric and sulphuric acids) and organic acids (e.g. acetic, citric or lactic acid) at or below pH 4.0. However, due to the deadly nature of the chlorine gas produced, handling is a primary limitation on the widespread use of chlorine dioxide.

Overall, the health concerns, tastes and odours and relatively high cost, due to its generation being on-site, and the concentrations that can be used in potable water to be effective, have tended to limit the uses of chlorine dioxide as a primary disinfectant for use in potable water. However, many water companies have been successfully using chlorine dioxide as a primary disinfectant, particularly where the water is above pH 8. It has been reported to cause problems of thyroid function and causing high serum cholesterol levels (Condie *et al.* 1994).

Mode of Action of Chlorine Dioxide

The mode of action of chlorine dioxide is primarily on the disruption of amino acids (Noss 1983) and RNA (Olivieri 1985) of Gram negative bacteria. In viruses the mode of action has been identified as disruption of the protein coat and the viral genome (Berman and Hoff 1984).

Effectiveness of Chlorine Dioxide on Biofilms

The legal requirement is that the combined concentration of chlorine dioxide, chlorite and chlorate should not exceed 0.5 mg l^{-1} chlorine dioxide equivalent. In order to determine that this concentration was able to control the presence of biofilms, and in particular *Legionella pneumophila*, a study was undertaken at the Building Services Research and Information Association (BSRIA). A full scale self contained rig was built to represent an office or residential building water services for 50 people (Pavey and Roper 1998).

The system was built in triplicate to allow thermal treatment to be compared with chlorine dioxide treatment in both hard and soft water. Sections of copper and glass reinforced plastic from the cold water storage tanks could be removed from the system to allow for analysis of biofouling before and during disinfection.

Results from the systems treated with chlorine dioxide demonstrated that control of *Legionella* within the biofilms took 20 days in the system that was using soft water and 60 days in the system using hard water (Walker *et al.* 1999b). This may indicate that the scaling that was occurring due to presence of the hard water may have been acting as a protective barrier and preventing the chlorine dioxide from working as efficiently as it did in the soft water. Other studies have shown that chlorine dioxide might kill oocysts of *Cryptosporidium parvum* in slightly contaminated water (Peeters *et al.* 1989) but there has been few studies to determine if the protozoan is controlled when enmeshed within biofilms (Rogers and Keevil 1995).

OZONE

Ozone is a pungent smelling and unstable gas. As a result of its instability, it is generated at point of use. Ozone generating apparatus includes a discharge electrode. To reduce corrosion, air is passed through a drying process and then into the ozone generator. The generator consists of 2 plates or a wire and tube with an electric potential of 15 000 to 20 000 volts. The oxygen in the air is dissociated by the impact of electrons from the discharge electrode. The atomic oxygen combines with atmospheric oxygen to form ozone.

$$O + O_2 \longrightarrow O_3$$

The resulting ozone–air mixture is then diffused into the water to be disinfected. It is a powerful oxidant known to be more powerful than hypochlorous acid. The advantage of ozone is that it does not form THMs. As with chlorine dioxide, ozone will not persist in water, decaying back to

O_2 in minutes. Therefore typically add ozone to raw water and filter for primary disinfection followed by chloramine addition after the filter as the distribution system disinfectant.

Ozone is very effective in potable water to remove taste, odour and colour, as the compounds responsible for these effects are unsaturated organics (Anselme *et al.* 1988). Ozone is seen as a very powerful disinfectant and is well known to be more effective than chlorine in the inactivation of *Giardia* cysts. Although ozone is not pH dependent its cidal activity decreases as the water temperature increases and so may have limited effect in hot water systems. However, one major drawback of using ozone is the fact that the residuals are quickly dissipated. The lifetime of ozone is usually less than an hour in most potable water systems (Glaze 1987). Due to this it is often necessary to use a secondary application of chlorine to provide disinfectant residual protection in potable water.

Mode of Action of Ozone

Ozone has been reported to disrupt bacterial membrane permeability, enzyme kinetics and also DNA (Ishizaki *et al.* 1984). It is also known to damage the nucleic acid core in viruses (Roy *et al.* 1981).

Effectiveness of Ozone on Biofilms

Ozone has been widely used in Europe and particularly in France as water disinfection in a number of water treatment plants (Miller 1978) with a 1–2 mg l^{-1} ozone dosage recommended for treatment of domestic water. In terms of treating biofilms, ozone has been used in the treatment of *L. pneumophila* on water fittings in hospitals. Although the *L. pneumophila* was eradicated from the fittings it was also removed from the control system that was ozone free, probably because this control system was subjected to other unforeseen treatments such as flushing and unexpected chlorine concentration increases (Edelstein *et al.* 1982).

ULTRAVIOLET LIGHT

The ultra-violet (UV) method of disinfection involves exposure of a thin layer of water to light from a mercury vapour arc lamp that emits UV. It has been found that submergence of the UV bulb encased in a quartz tube is superior to overhead irradiation. The depth of light penetration still limits the liquid film thickness around each lamp to about 50–80 mm. Multiple lamps are used to provide greater coverage, the major factor in achieving good microbial kill is the ability of the UV light to pass through water to get to the target organisms. Thus lamps must be kept free of slime and precipitates and the water must be free of turbidity. UV performs well against bacteria and viruses. Major disadvantages are that it leaves no residual protection for the distribution system and it is very explosive.

UV disinfection was first used at the beginning of the Century to treat water in Kentucky, but it was abandoned in favour of chlorination. Owing to technological improvements this alternative disinfection mechanism is now regaining popularity, particularly in Europe (Wolfe 1990). UV disinfection systems use low pressure mercury lamps enclosed in quartz tubes. The tubes are immersed in flowing water in a tank and allow passage of UV radiation at germicidal wavelengths of 2537 A. However, transmission of UV by quartz decreases upon continuous use. Therefore, the quartz lamps must be regularly cleaned by mechanical, chemical and ultrasonic cleaning methods. Teflon has been proposed as an alternative to quartz but its transmission of UV radiation is lower than in the quartz systems.

Mechanisms of Action of Ultraviolet Light

Studies with viruses have demonstrated that the initial site of UV damage is the viral genome, followed by structural damage to the virus coat (Rodgers et al. 1985). UV radiation damages microbial DNA at a wavelength of approx. 260 nm. It causes thymine dimerization, which blocks DNA replication and effectively inactivates microbes.

Microbial inactivation is proportional to the UV dose, which is expressed in microwatt—seconds per sq. cm. The inactivation of microbes by UV radiation can be expressed by the following equation (Luckiesh and Holladay 1944):

$$N/No = e^{-kpdt}$$

No = initial number of microorganisms (/ml); N = number of surviving microbes (/ml); k = inactivation rate constant (μW s cm^{-2}); Pd = UV light intensity reaching the organisms (μW cm^{-2}); and t = exposure to time in seconds.

The above equation is subject to several assumptions, one of which is the log of the survival fraction should be linear with regard to time (Severin 1980). In environmental samples, however the inactivation kinetics are not linear with time, which may be due to resistant organisms among the natural population and to differences in flow patterns.

The efficacy of UV disinfection depends on the type of microorganisms under consideration. In general the resistance of microbes to UV follows the same pattern as with chemical disinfectants, which is as follows (Chang et al. 1985): protozoan cysts > bacterial spores > viruses > vegetative bacteria. This trend was supported by other researchers that showed the approximate

UV dose (μW s cm^{-2}) for 90% inactivation of microbes (Wolfe 1990). A virus such as hepatitis A requires a UV dose of 2700 (μW s cm^{-2}) for one-log inactivation but necessitates 20 000 mW sec cm^{-2} to achieve a 3 log reduction (Wolfe 1990).

Many variables (e.g. suspended particles, chemical oxygen demand, colour) in waste water effluents affect UV transmission in water and thus disinfection efficiency (Harris *et al.* 1987; Severin 1980). Several organic compounds (e.g. humic substances, phenolic compounds, lignin sulphonates from pulp and paper mill industry, ferric iron) interfere with UV transmission in water. Indicator bacteria are partially protected from the harmful UV radiation when embedded within particulate matter (Oliver and Cosgrove 1977; Qualls *et al.* 1983; Qualls *et al.* 1985; Harris *et al.* 1987). Suspended solids protect microorganisms only partially from the lethal effect of UV radiation. This is because suspended particles in water and wastewater absorb only a portion of the UV light (Bitton *et al.* 1972).

Effectiveness of Ultraviolet Light on Biofilms

One major advantage of UV disinfection is that it is able to destroy microbial life in the water phase without the addition of anything to the water. However when applied to the control of biofilms this characteristic is also a disadvantage as UV disinfection leaves no residual. Hence UV disinfection can control for example the incoming source water thus supplying in essence sterile water and prevent biofilm formation. There will be no distribution system or network that will remain sterile from assembly and commissioning, and so although UV may help to maintain the cleanness of an already sterile system, an additional chemical disinfectant such as chlorine or bromine is added post UV disinfection.

IONIZATION

The process of ionization has been well documented (Sykes 1965) over many years and is based upon electrolysis in which ions undergo electron transfer at an electrode surface. In water services the technique is concerned with releasing silver and copper ions into the water by passing an electrical current between electrodes placed in running water. As the electrons pass between the anode (+ve) and cathode (−ve), one or two electrons are left behind on the anode surface. As the remaining electrons travel across to the cathode they are driven away by the flow of the water into solution. These ions in solution represent charged atoms or groups of atoms where one or more electrons have been lost and the atom is no longer neutral but carries a charge. In the case of silver ions these are designated Ag$^+$ where one electron is missing and the ions carry a single positive charge. In the case of copper ions these are designated Cu^+ or Cu^{2+} depending upon whether one or two electrons are missing. Ionization units are used on location and, in general, consist of an electrode chamber and control unit. Typically the chamber will contain silver/copper alloy electrodes of between 10–30% silver, dependent on manufacturer. The size and number of electrodes will be dependent upon the type of application according to the water volume, flow rate and required microbial control. In a number of studies, the combination of these metals with halogenation has been shown to have applications in the disinfection of both recreational and potable water (Pyle *et al.* 1992).

Mode of Action of Ionization

Ionization has been used in the control of both water borne bacteria (Landeen *et al.* 1989) and viruses (Abad *et al.* 1994) over many years. Copper ions kill bacteria by destroying cellular protein due to the oxidation of sulphydryl groups of enzymes thus interfering with respiration (Hugo and Russel 1982; Domek *et al.* 1984). Silver ions also interfere with enzyme activity by binding to proteins whilst both ions bind to DNA molecules. Advantages of ionization are that a residual is maintained throughout the systems.

Effectiveness of Ionization on Biofilms

The effectiveness of ionization on water borne microorganism has been well proven to be successful (Pyle *et al.* 1992). The use of this technology against biofilms has been studied, particularly against *L. pneumophila* (Mietzner *et al.* 1997; Rohr *et al.* 1996; Rogers *et al.* 1995; Walker *et al.* 1999b). However there are problems with using this technology in the field, where parameters cannot always be guaranteed to remain constant. In one study where ionization was compared in soft and hard water, complications were due to the scaling up of the electrodes and drifting pH of the water, leading to failure to control *L. pneumophila* (Pavey 1996; Walker *et al.* 1999a). Although these problems were rectified it demonstrates the inherent problems of controlling pathogens with automated disinfection processes that are susceptible to changes in water chemistry.

OTHER BIOCIDES USED IN POTABLE WATER

Potassium permanganate is often used in water supply treatment particularly for the removal of, taste and odour and metal ions, such as iron and manganese (Cherry 1962; Shull 1962). It has limited disinfection efficacy and is not as effective as the use of chlorinated water (Cleasby *et al.* 1964; Buelow *et al.* 1976). There is a possible benefit of using potassium permanganate as a peroxidant in the early stages of the treatment process as it is well known to reduce the growth of algae and slime bacteria.

FUTURE METHODS IN THE CONTROL AND REMOVAL OF BIOFILMS

Methods have been suggested that may be developed to release bacteria from surfaces (Brisou 1995). This release, using enzymes can act: (i) directly on microbial adhesins; (ii) on the structures of the media-sensitive enzymes; (iii) on the polysaccharides produced by the bacteria during surface colonization and (iv) on aggregates containing the bacteria.

It has been shown that hydrolases can release bacteria from surfaces with exposure times of 2–4 h (Brisou 1995). Enzymes such as hydrolases have been shown to free oligosaccharides and monosaccharides. If these products could be applied to biofilms within potable water environments it may release bacteria from the surface of potable water pipes enabling greater disinfection. Could this be an alternative to the use of biocides in the future? If this is both practical and feasible more work is needed in this area to enable a better understanding of the processes involved, and to provide solutions to problems of biofilms. However, the diverse range of bacteria found within the biofilm and the complexity of extracellular polymeric substances found make it a very hard and daunting task as an alternative short term solution.

DISINFECTANT RESISTANT ORGANISMS

Different bacteria and viruses vary in their resistance to disinfectants. Particular factors relevant to this are that of pH, temperature, disinfectant concentration and contact time, as changing any one of these conditions will produce different rates of inactivation for the same organism. The major factor to consider during the conduct of any disinfection regime is that: viruses and protozoa are more resistant to disinfection than enteric bacteria.

It is generally found that heterotrophic bacterial populations can be controlled to levels of 500 organisms per millilitre in many water supplies by the addition and maintenance of 0.3 mg l^{-1} residual chlorine (Geldreich *et al.* 1996). From this work it was found that any further increases in the residual chlorine concentration did not result in any significant decreases in the heterotrophic bacterial densities. The reason for this was that organisms were being protected in sediment habitats, while selective pressures induced the growth of resistant organisms.

Within studies carried out on chlorine resistance of bacteria (Ridgway and Olson 1982) present in potable water the greatest resistance has been observed in Gram positive, spore forming bacilli, *Actinomycetes* and some micrococci (Ridgway and Olson 1982). These organisms were found to survive 2 min exposures to 10 mg l^{-1} free chlorine. In contrast it was found that organisms most sensitive to chlorine contact were *Corynebacterium/ Arthrobacter*, *Klebsiella*, *Pseudomonas/Alcaligenes*, *Flavobacterium/Moraxella* and *Acinetobacter*. It was found in this study that these organisms were inactivated by 10 mg l^{-1} or less of free chlorine.

OVERALL CONTROL OF BIOFILMS IN GENERAL

The actual elimination of biofilms is very difficult and a very frustrating process for engineers and biologists. Mechanical cleaning is seen as the most efficient way of removing biofilms. However, the actual design and construction of the equipment being cleaned compromises the cleaning process. This is particularly so with respect to methods employed such as brush balls for circulating systems and high pressure washing systems because they cannot be used for closed systems. Cleaning in place systems have not been designed to eliminate biofilms although this method can reduce biofilm development if the equipment design and materials are suitable.

The choice of materials and their surface topographies are very important particularly with respect to cleaning and preventing biofilm formation. Any system should be designed without deadlegs and should be cleaned regularly to avoid the accumulation of biofilm. The cleanliness of surfaces, training of personnel and good manufacturing and design practice are the most important approaches in combating biofilm formation.

Whilst biocides are used extensively within both cooling towers and within the pulp and paper industry it should be borne in mind that oxidants and other chemicals added for specific disinfectant processes may actually increase the amount of biodegradable material in treated water. Therefore, biocides which are designed to prevent biofilm development actually increase the amount of nutrients available for growth. It should also be borne in mind that the use of biocides is very important considering the complexity of biofilms in that they consist of a complex microbial community and copious amounts of extracellular polymeric material. As biofilms have been demonstrated to provide a safe haven for maintaining the viability of the pathogens the disruption of the polymeric matrix will assist the delivery of the disinfectant to the microbes (Walker *et al.* 1995). This EPS material may also act as a nutrient source in particular under extreme starvation conditions.

The existence of biofilms in a potable water system should always be considered as a potential haven for the proliferation of pathogens, in particular coliforms. The presence of pathogens suggest that there are deficiencies within the potable water system constituting a public health threat. The existence of these organisms within the biofilm is due to many factors possibly treatment breakthrough from sewage pipes or cross connection problems. When we consider the development of coliform aftergrowth within a biofilm the best measure for control is that of anticipation. The factors known to contribute to the development of a biofilm are numerous. However, providing that the growth of a biofilm can be limited in potable water this should ensure biofilm growth is limited.

SHORT TERM CONTROL OF BIOFILMS

Prevention methods within areas where biofouling is evident involve regular cleaning but this does not prevent viable bacteria re-colonizing the surface. Physical scouring or 'pigging' helps control biofilm build up if combined with organic acids or alkalis. Other attempts include filtration devices, which are quickly fouled, or UV radiation. Transmission of UV radiation however, is decreased in turbid water and has poor penetration in microflocculations. Heating water systems to perhaps 70° C for one hour may act as a controlling parameter for biofouling (Geesey et al. 1994), but care has to be taken to avoid possible scalding. Chlorination seems to be the most practicable method used for disinfection of potable water supplies compared to monochloramines because residual levels can readily be measured. Chlorine is, however, sensitive to temperature and light, inactivated by organic matter and its efficacy as a bacterial disinfectant decreases rapidly at pH values over 7.

In practice, the maintenance of an effective free chlorine residual concentration in a water system cannot be relied upon to prevent biofilm formation. A range of heterotrophs have been recovered from water containing concentrations of free chlorine of 0.1–0.5 ppm. There are two reasons for this. In a fast flowing pipe, there is always a thin layer of slower moving water (the viscous sublayer) just above the biofilm. Any disinfectants have to pass through this layer. Free chlorine is highly reactive but it is not persistent so its ability to affect biofilms is reduced by the presence of the viscous sublayer. Chloramines are less reactive than free chlorine but are more persistent and are able to penetrate the laminar layer to a greater extent. Chlorine-based disinfectants are unable to penetrate deep into the biofilm because of its polymer gel structure.

Classical disinfectants, as mentioned above (e.g. chlorine or chloramines) are ineffective to control attached biomass so it is necessary to control attached biomass using several techniques such as limiting biologically degradable organic carbon (BDOC) and the concentration of suspended bacterial cells in the water entering the distribution system. Biofilm control in a distribution system is complicated and requires continuous action. The characteristics of the finished waters feeding the system have to be carefully controlled (low BDOC, low cell concentration). A secondary chlorination (i.e. chlorination of water already in the distribution system) is not a curative treatment but an additional precaution for killing planktonic microorganisms; its efficiency is directly related to the previous organic matter reduction and a good hydraulic regime.

Ultimately limiting the BDOC and the concentration of planktonic microbial cells entering any potable water system would provide a better means of control. Studies using 3–5 mg l^{-1} of active chlorine have demonstrated that this concentration can be effecive for eliminating biofilms (Nagy et al. 1982). However it has been reported that even 10 mg l^{-1} of chlorine was not sufficient for biofilm control (Exner et al. 1987). The efficiency of any biocide as a means of disinfection will ultimately depend on BDOC levels of the water, nutrient levels within the biofilm, the age of the biofilm, the surface material and amount of extracellular material present (LeChevallier et al. 1988b). Within potable water systems, chloramine-based compounds (monochloramine) seem to be growing in use because of improved penetration into biofilms (LeChevallier et al. 1988b). It has been found that polysaccharides that constitute the matrix of the biofilm can be easily penetrated suggesting that the age or characteristics of the microorganisms within the biofilm are not important for biocidal efficacy (LeChevallier et al. 1991). Whilst the effects of chloroamine have been found not to be as effective as hypochlorite at reducing planktonic microbes, its effectiveness in penetrating the exopolymer matrix makes it a better candidate for use in biofilm removal. Monochloramine was the most effective at killing sessile bacteria growing on a solid metallic surface when compared to hypochlorous acid and chlorine dioxide (LeChevallier et al. 1991).

It is generally found that if a number of control measures are used for removing biofilms then recurrence of the problem begins again after only one week (LeChevallier et al. 1987).

LONG TERM CONTROL OF BIOFILMS IN POTABLE WATER

Biofilm growth seems to be difficult to stop but there are several ways biofilms can be controlled. These include: (i) reduction of nutrient concentrations in rivers and lakes, which would reduce the potential for biofilm growth in potable water systems, as would nutrient reduction at the water treatment works. Without nutrients, biofilms are not able to thrive and mature. Therefore starving bacteria by removing nutrients from the potable water may eventually be the solution to control the presence of the microbial biofilm, however, it currently involves expensive technologies; (ii) using materials in potable water systems which do not leach nutrients, thus reducing excessive biofilm formation. In the U.K., non-metallic materials in contact with potable water must comply to BS 6920; (iii) effective management of the hydraulics of distribution systems to avoid slow-moving or stagnant pockets of water will help control biofilm maturation, and the continuous presence of disinfectant residual has a suppressive effect; (iv) the treatment of source water according to internationally approved standards which will destroy pathogenic organisms. However, distribution and plumbing systems can be colonized with microbes. The appearance of faecal organisms may be due to their surviving disinfection because they may be protected within biofilms. The metabolic activity within a biofilm may protect species sensitive to changes in pH or high oxygen tension of the circulating water. It has been shown that in a potable water distribution system, even in the absence of chlorine, bacterial growth in the liquid phase is negligible (LeChevallier *et al.* 1987). Essentially only the bacteria in biofilm attached to the walls of the distribution pipeline are multiplying and, due to shear loss, constitute one of the main causes of deterioration of microbiological quality of water distribution systems.

CONCLUSION

Biofilm control within potable water systems is complicated and requires immediate action with respect to potential waterborne disease implications. The major control with respect to reducing biofilm accumulation is governed by the careful control of water BDOC levels and maintaining but ultimately reducing, cell count levels. Post disinfection with the use of chlorine is by no means a curative measure but a precautionary measure when biofilm growth and coliform aftergrowth is evident. At present, biofilm development within potable water cannot be completely controlled and therefore an associated problem is that of public health significance due to the presence of pathogens. Also the increasing isolation of bacteria that are resistant to the disinfectant concentration will only complicate the argument. Even if increased chlorine concentration is the answer, the development of secondary precursors will have a major long term effect on human health. It is not possible to fully assess the performance of disinfection on biofilms in potable water distribution systems due to the constantly changing number of variables. Whilst these have been looked at in many pilot and laboratory-based experiments they have lead to a number of conflicting results and unanswered questions. Therefore it is evident that the issue of biofilm control is very complex and requires carefully planned monitoring and maintenance programmes to follow the progression of the biofilm and the effect of control measures that are applied.

REFERENCES

Abad, FX, Pinto, RM, Diez, JM and Bosch, A (1994) Disinfection of human enteric viruses in water by copper and silver in combination with low levels of chlorine. *Applied and Environmental Microbiology* 60:2377–2383.

Anselme, C, Suffet, IH and Mallevialle, J (1988) Effects of ozonation on tastes and odors. *Journal of American Water Works Association* 80:45–51.

Berman, D and Hoff, JC (1984) Inactivation of simian rotavirus SA11 by chlorine, chlorine dioxide, and monochloramine. *Applied and Environmental Microbiology* 48:317–323.

Berman, D, Rice, EW and Hoff, JC (1988) Inactivation of particle-associated coliforms by chlorine and monochloramine. *Applied Environmental Microbiology* 54:507–512.

Bitton, G (1994) Wastewater Microbiology, Wiley-Liss, York.

Bitton, G, Henis, Y and Lahav, N (1972) Effect of several clay minerals and humic acid on the survival of *Klebsiella aerogenes* exposed to ultraviolet irradiation. *Applied and Environmental Microbiology* 23:870–874.

Brisou, JF (1995) Biofilms—methods for enzymatic release of microorganisms, CRC Press, New York.

Buelow, RW, Taylor, RH, Geldreich, EE, Goodenkauf, A, Wilwerding, L, Holdren, F, Hutchinson, M and Nelson, IH (1976) Disinfection of New Water Mains. *Journal of the American Water Works Association* 68:283–288.

Bull, RJ (1982) Health effects of drinking water disinfectants and disinfectant by-products. *Environmental Science and Technology* 16:554A.

Chang, JC, Ossoff, SF, Lobe, DC, Dorfman, MH, Dumais, CM, Qualls, RG and Johnson, JD (1985) UV inactivation of pathogenic and indicator microorganisms. *Applied and Environmental Microbiology* 49:1361–1365.

Characklis, WG (1981) Fouling biofilm development: a process analysis. *Biotechnology and Bioengineering* 23:1923–1960.

Characklis, WG and Marshall, KE (1990) Biofilms, John Wiley & Sons Inc, New York.

Cherry, AK (1962) Use of potassium permanganate in water treatment. *Journal of the American Water Works Association* 54:417–424.

Cleasby, JL, Bauman, ER and Black, CD (1964) Effectiveness of potassium permanganate for disinfection. *Journal of American Water Works Association* 56:466–474.

Condie, LW, Lauer, WC, Wolfe, GW, Czeh, ET and Burns, JM (1994) Denver Potable Water Reuse Demonstration Project: comprehensive chronic rat study. *Food Chemistry and Toxicology* 32:1021–1030.

Domek, MJ, LeChevallier, MW, Cameron, SC and McFeters, GA (1984) Evidence for the role of copper in the injury process of coliform bacteria in drinking water. *Applied and Environmental Microbiology* 48:289–293.

Edelstein, PH, Whittaker, RE, Kreiling, RL and Howell, CL (1982) Efficacy of ozone in eradication of *Legionella pneumophila* from hospital plumbing fixtures. *Applied and Environmental Microbiology* 44:1330–1333.

Exner, M, Tuschewitzki, GJ and Scharnagel, J (1987) Influence of biofilms by chemical disinfectants and mechanical cleaning. *Zentralbl Bakteriol Mikrobiol Hyg [B]* **183**:549–563.

Geesey, GG, Bremer, PJ, Fischer, WR, Wagner, D, Keevil, CW, Walker, JT, Chamberlain, AHL and Angell, P (1994) Unusual types of pitting corrosion of copper tubes in potable water systems In: Biofouling and Biocorrosion in Industrial Water Systems (Eds. Geesey, GG, Lewandowski, Z and Flemming, H-C), Lewis Publishing Company, London, pp. 243–263.

Geldreich, EE (1996) Microbial Quality of water supply in distribution systems, Lewis, New York.

Glaze, WH (1987) Drinking water treatment with ozone. *Environmental Science and Technology* **21**:224–230.

Harris, GD, Adams, VD, Sorensen, DL and Dupont, DR (1987) The influence of photoreactivation and water quality on ultraviolet disinfection of secondary municipal wastewater. *Journal of the Water Pollution Control Federation* **59**: 781–787.

Hugo, WB and Russel, AD (1982) Historial Introduction. In: Principles and practices of disinfection, preservation and sterilisation (Eds. Russel, AD, Hugo, WB and Aycliffe, GAJ), Blackwell, Oxford, pp. 8–106.

Ishizaki, K, Shinriki, N and Ueda, T (1984) Degradation of nucleic acids with ozone. V. Mechanism of action of ozone on deoxyribonucleoside 5′-monophosphates. *Chemical Pharmacological Bulletin (Tokyo)* **32**:3601–3606.

Jacangelo, JG and Olivieri, VP (1985) Aspects of the mode of action of mono-chloramine. In: Water Chlorination, Chemistry, Environmental Impact and Health Effects (Eds. Jolley, RL, Bull, RJ, Davis, WP, Katz, S and Roberts, MH), Lewis Publishers, Chelsea, Mich.

Knox-Holmes, B (1993) Biofouling control with low levels of copper and chlorine. *Biofouling* **7**:157–166.

Landeen, LK, Yahya, MT and Gerba, CP (1989) Efficacy of copper and silver ions and reduced levels of free chlorine in inactivation of *Legionella pneumophila*. *Applied and Environmental Microbiology* **55**:3045–3050.

LeChevallier, MW, Babcock, TM and Lee, RG (1987) Examination and characterization of distribution system biofilms. *Applied and Environmental Microbiology* **53**:2714–2724.

LeChevallier, MW, Cawthon, CD and Lee, RG (1988a) Factors promoting survival of bacteria in chlorinated water supplies. *Applied and Environmental Microbiology* **54**:649–654.

LeChevallier, MW, Cawthon, CD and Lee, RG (1988b) Inactivation of biofilm bacteria. *Applied and Environmental Microbiology* **54**:2492–2499.

LeChevallier, MW, Hassenauer, TS, Camper, AK and McFeters, GA (1984) Disinfection of bacteria attached to granular activated carbon. *Applied and Environmental Microbiology* **48**:918–923.

LeChevallier, MW, Schulz, W and Lee, RG (1991) Bacterial nutrients in drinking water. *Applied and Environmental Microbiology* **57**:857–862.

Luckiesh, M and Holladay, LL (1944) Disinfecting water by means of germicidal lamps. *General Electrical Review* **47**:45–54.

Malpas, JF (1973) Disinfection of water using chlorine dioxide. *Water Treatment Examinations* **22**:209–221.

Mietzner, S, Schwille, RC, Farley, A, Wald, ER, Ge, JH, States, SJ, Libert, T, Wadowsky, RM and Miuetzner, S (1997) Efficacy of thermal treatment and copper-silver ionization for controlling *Legionella pneumophila* in high-volume hot water plumbing systems in hospitals. *American Journal of Infectious Control* **25**:452–457.

Miller, GW (1978) Environmental Protection Series, pp. EPA-600/2-78-147.

Montgomery, JM (1985) Water Treatment Principles and Design, John Wiley & Sons, New York.

Nagy, LA, Kelly, AJ, Thun, MA and Olson, BH (1982) Biofilm composition, formation and control in the Los Angeles aqueduct system. Proceedings of the Water Quality Technical Conference, Nashville, TN.

Neden, DG, Jones, RJ, Smith, JR, Kireyer, GJ and Foust, GW (1992) Comparing chlorination and chloramines for controlling bacterial regrowth. *Journal of the Amercian Water Works Association* **84**:80.

Noss, CI (1983) Reactivity of chlorine dioxide with nucleic acids and proteins. In: Water chlorination: environmental impacts and health effects, Vol. 4 (Ed. Jolley, RJ), Ann Arbor Scientific, Michigan.

Oliver, BG and Cosgrove, EG (1977) The disinfection of sewage treatment plant effluents using ultraviolet light. *Canadian Journal of Chemical Engineering* **53**:170–174.

Olivieri, VP (1985) Mode of action of chlorine dioxide on selected viruses. In: Water Chlorination:environmental impact and health effects, Vol. 5 (Ed. Jolley, RJ), Michigan.

O'Neill, JG, Banks, J and Jess, JA (1997) Biofilms in water mains—now under control. In: Biofilms: Community interactions and control (Eds. Wimpenny, J, Handley, P, Gilbert, P, Lappin-Scott, H and Jones, M), Bioline, Cardiff, pp. 259–268.

Pavey, N (1996) Ionisation water treatment—for hot and cold water services, Bourne Press, Bracknell.

Pavey, NL and Roper, M (1998) *Chlorine dioxide water treatment—for hot and cold water services*, Oakdale Printing Co, Surrey.

Peeters, JE, Mazas, EA, Masschelein, WJ, Villacorta Martiez de Maturana, I and Debacker, E (1989) Effect of disinfection of drinking water with ozone or chlorine dioxide on survival of *Cryptosporidium parvum* oocysts. *Applied and Environmental Microbiology* **55**:1519–1522.

Percival, AL, Beech, IB, Knapp, JS, Wales, DS and Edyvean, RG (1997) Biofilm development on stainless steel in a potable water system. *Journal of the Institute of Water and Environmental Management* **14**:289–294.

Pyle, BH, Broadaway, SC and McFeters, GA (1992) Efficacy of copper and silver ions with iodine in the inactivation of *Pseudomonas cepacia. Journal of Applied Bacteriology* **72**:71–79.

Qualls, RG, Flynn, MP and Johnson, JD (1983) The role of suspended particles in ultraviolet irradiation. *Journal of the Water Pollution Control Federation* **55**:1280–1285.

Qualls, RG, Ossoff, SF, Chang, CH, Dorfman, MH, Dumais, CM, Lobe, DC and Johnson, JD (1985) Factors controlling sensitivity in ultraviolet disinfection of secondary effluents. *Journal of the Water Pollution Control Federation* **57**:972–987.

Ridgway, HF and Olson, BH (1982) Chlorine resistance patterns of bacteria from two drinking water distribution systems. *Applied and Environmental Microbiology* **44**:972–987.

Rodgers, FG, Hufton, P, Kurzawska, E, Molloy, C and Morgan, S (1985) Morphological response of human rotavirus to ultra-violet radiation, heat and disinfectants. *Journal of Medical Microbiology* **20**:123–130.

Rogers, J, Dowsett, AB and Keevil, CW (1995) A paint incorporating silver to control mixed biofilms containing *Legionella pneumophila. Journal of Industrial Microbiology* **15**:377–383.

Rogers, J and Keevil, CW (1995) Survival of *Cryptosporidium parvum* in aquatic biofilms. In: Protozoal Parasites in Water, Vol. 58 (Eds. Thompson, KC and Fricker, C), Royal Society for Chemistry, London, pp. 209–213.

Rohr, U, Senger, M and Selenka, F (1996) Effect of silver and copper ions on survival of *Legionella pneumophila* in tap water. *Zentralbl Hyg Umweltmed* **198**:514–521.

Roy, D, Wong, PK, Engelbrecht, RS and Chian, ES (1981) Mechanism of enteroviral inactivation by ozone. *Applied and Environmental Microbiology* **41**:718–723.

Severin, BF (1980) Disinfection of municipal waste water effluents with ultraviolet light. *Journal of the Water Pollution control Federation* **52**:2007–2018.

Shull, KE (1962) Operating experiences at Philadelphia suburban treatment plants. *Journal of the American Water Works Association* **54**:1232–1240.

Sykes, G (1965) The Halogens. In: Disinfection and Sterilisation. Chapman and Hall, London, pp. 381–410.

Tanner, RS (1989) Comparative testing and evaluation of hard-surface disinfectants. *Journal of Industrial Microbiology* **4**:145–154.

Walker, JT, Ives, S, Morales, M and West, AA (1999a) Control and monitoring of biofouling using an avirulent *Legionella pneumophila* in a water system treated with silver and copper ions. In: Biofilms in Aquatic Systems (Eds. Keevil, CW, Godfree, A, Holt, D and Dow, C), Society for Applied Microbiology, London, pp. 131–138.

Walker, JT, Mackerness, CW, Rogers, J and Keevil, CW (1995) Biofilm—A haven for Waterborne pathogens. In: Microbial Biofilms (Eds. Lappin-Scott, HM and Costerton, JW), Cambridge University Press, London, pp. 196–204.

Walker, JT, Roberts, ADG, Lucas, VJ, M Roper, M and Brown, R (1999b) Quantitative assessment of biocide control of biofilms and legionella using total viable counts, fluorescent microscopy and image anaylsis. *Methods in Enzymology,* **310**, 629–637.

Wolfe, R. L. (1990) Ultraviolet disinfection of potable water. *Environmental Science and Technology* **24**:768–773.

3 Biofouling of Industrial Waters and Processes

3.1 Problems of Biofilms in Industrial Waters and Processes

GLYN MORTON
Department of Biological Sciences,
University of Central Lancashire, Preston, Lancashire, UK

INTRODUCTION

Fouling has been defined as the undesirable formation of inorganic and/ or organic deposits on surfaces, whilst biological fouling or biofouling is regarded as the accumulation and growth of living organisms and their associated organic and inorganic material on a surface (Edyvean and Videla 1994). Microorganisms have a straightforward approach to life; they use whatever is available as a food source, attach themselves to practically all surfaces, multiply and build up biomass (Heitz *et al.* 1996). This surface associated microbial activity and colonization, or biofilm formation is a phenomenon that occurs in both natural and man-made environments, especially in nutrient limited conditions. The cell walls of Gram-positive bacteria, the outer membranes of Gram-negative bacteria and their capsules, all contain polysaccharides in various amounts (Gaylarde and Beech 1989). Polysaccharides in the form of lipopolysaccharides (LPS) form side chains, which project from Gram-negative bacterial cell surfaces and form the interface between other adjacent cells and/or the surrounding environment (Peterson and Quie 1981). LPS may selectively bind extracellular cations. This chelating property has been implicated in the biocorrosion of metals (Beech and Gaylarde 1991). The initial conditioning of the surface of a metal by inorganic and organic molecules is followed by the adhesion of bacterial cells (Videla 1990). Extracellular polysaccharide substance (EPS) which is secreted from some bacterial cells, may enhance the adhesion of these cells and of those in the immediate vicinity leading to the formation of a biofilm (Beech and Gaylarde 1991). EPS has also been implicated in chelation of metal ions (Lieve *et al.* 1968; Ferris *et al.* 1987). Biofilms effect the interaction between surfaces and the environment, not only in biodeterioration processes like corrosion but also in several biotechnological processes (Videla and

Industrial Biofouling. Edited by J. Walker, S. Surman, J. Jass
©2000 John Wiley & Sons Ltd

Characklis 1992). Biofilms exist as beneficial epilithic communities in rivers and streams and they play an important role in waste water treatment plants on trickling filter beds (Costerton *et al.* 1986; Bryers and Characklis 1982). However they are better known for their widespread deleterious effects (Table 3.1.1).

Table 3.1.1. Examples of the detrimental effects of biofilms

Occurrence	Problem/Effect	Reference
Ship hull fouling	Increased fuel cost	Zobell and Allen 1935; Dempsey 1981a; Dempsey 1981b
Fluid flow systems	Energy loss and equipment failure, reduced heat exchange capacity	McCoy *et al.* 1981; Elsmore and Dorman 1988; Brankevich *et al.* 1990; Rittenhouse 1991
Water distribution	Organoleptic problems, regrowth and contamination with potential pathogens, decreased pipeline capacity, pitting corrosion	LeChevallier *et al.* 1988a; LeChevallier *et al.* 1988b; van der Wende *et al.* 1989; Exner *et al.* 1987; Shariff and Hassan 1985; Walker *et al.* 1991
Desalination	Membrane flux decline, membrane degradation, increased salt passage	Abd El Aleem *et al.* 1998
Oil recovery industry	Corrosion brought about by sulphate reducing bacteria and blockage of pipelines	Lynch and Edyvean 1988; Shaw *et al.* 1985; Herbert 1994
Hydrocarbon fuels	Contamination of diesel fuels	Parberry 1971; Cofone *et al.* 1973
Food, paper and paint industries	Spoilage and decreases in product quality	Heaton *et al.* 1991; Holah *et al.* 1994; Eastwood 1994; Väisänen *et al.* 1994
Metal working fluid industry	Breakdown of emulsions, blocking of pipes and filters, presence of pathogens	Rossmore 1986; Hill 1975; Herwaldt *et al.* 1984; Prince and Morton 1988
Attachment to medical devices and prosthecae	Resistance to antimicrobial treatments and increased risk of secondary infection	Gristina *et al.* 1985; Costerton *et al.* 1987; Anwar *et al.* 1989; Kristinsson 1989; Anwar and Strap 1992; Reid and Busscher 1992
Attachment to teeth	Dental caries and plaque formation	Keevil *et al.* 1987; Marsh *et al.* 1993

Modified from Morton *et al.* (1998).

BIOFOULING AND BIODETERIORATION

Biodeterioration may be defined as 'the deterioration of materials of economic importance by microorganisms' (Hueck 1965). Biodeterioration is due to any undesirable change in the properties of a material caused by the vital activities of organisms, including microorganisms. Biodeterioration can be brought about by mechanical processes, where the material is damaged as a direct result of the activity of an organism, such as its movement or growth. Chemical dissimilatory processes (perhaps the most common form of biodeterioration), occur when a material is degraded for its nutritive value, whilst chemical assimilatory processes occur when metabolic products damage a material by causing corrosion, pigmentation or the release of toxic metabolites. Soiling and biofouling occur when the mere presence of an organism or its excrement renders the product unacceptable. As a result of biofouling, one or more, or indeed all of these deleterious processes may occur.

BIOFOULING OF INDUSTRIAL WATERS

Any untreated water used in an industrial process may carry microorganisms. In many industries the formation of biofilms within pipework, cooling systems, heat exchangers and filters can cause problems. The resulting losses of efficiency due to increased frictional resistance in pipes (McCoy *et al.* 1981) or decrease in heat exchange capabilities (Trulear and Characklis 1982; Shariff and Hassan 1985) can result in decreased production rates and increased costs. Characklis and Cooksey (1983) provide us with information on the effects and relevance of biofilms on various rate processes. These include, heat transfer reduction, increase in fluid frictional resistance, mass transfer and chemical transformations.

WATER SYSTEMS

The incidence of biofilms within industrial water distribution systems is well documented (Costerton *et al.* 1987; Le Chevallier *et al.* 1988; Colbourne *et al.* 1988). There are two main types of water systems. Open or once through, which tend to handle large volumes of water e.g. potable water supplies, sewers and industrial systems where water is used in large quantities, and closed systems which deal with limited volumes of water (or water based commercial fluid) e.g. recirculating water cooling systems and metal working fluid systems (see later). The addition of biocides and corrosion inhibitors to open systems is not practicable (Edyvean 1990).

COOLING TOWERS

Modern industrial cooling towers are built to different designs according to particular site specifications and requirements (Eaton 1976). They act as large heat exchangers by allowing warm water produced from an industrial plant to be sprinkled over packing or filling material in the tower and then recirculated back to the plant (Eaton 1994). Cooling towers readily generate fine water droplets as they distribute water over packing material through which there is a counter-current flow of air. Cooling tower water, make-up water and effluent form special aquatic habitats with their own physico-chemical characteristics (Udaiyan and Manian 1991). Wood is widely used as the packing material and the warm (25–35° C) wet environment provides ideal conditions for the proliferation of microorganisms. Savory (1954) reports soft rot decay by members of the ascomycetes and fungi imperfecti, while more recently Singh *et al.* (1992) and Eaton (1994) record soft rot fungi and tunnelling bacteria as the culprits. The preferred treatment of cooling tower timber packing in the UK has been pressure impregnation of the wood with copper chrome arsenic (CCA) preservatives. The constant flow of water over the surfaces also promotes leaching of non-fixed preservative from the wood. Figure 3.1.1, shows a sectional view of a mechanical–induced–draught cooling tower.

The environment of the air conditioning cooling tower has also proved to be an ideal environment for the growth of biofilms harbouring a range of aquatic microorganisms, including legionellae. *Legionella pneumophila*, the aetiological agent responsible for legionnaires' disease, is a microorganism which is widespread within the environment. It is an opportunistic human pathogen found in high numbers in both natural and man-made aquatic environments (Grimes 1991). Cooling towers and evaporative condensors are cited as potential sources of the pathogen by Tobin *et al.* (1981), Breiman (1993), and Bentham (1993), who reports colonization by *Legionella* spp. in 31 cooling towers in South Australia. The sloughing off or erosion of biofilms may also liberate *Cryptosporidia* and *Giardia* spp. into the water (Reasoner 1988).

BIOFILMS AND PIPELINES

The two main functions of pipelines in oilfield operations are to carry water, water for injection into reservoirs to maintain pressure, and water to carry the produced oil, from the wells to treatment, and then to storage and export facilities. The pipelines may be constructed of mild or stainless steels and in some cases plastic pipes may be used (Herbert 1994). The diameters of the pipes may range from a few inches to a few feet, whilst the pipelines

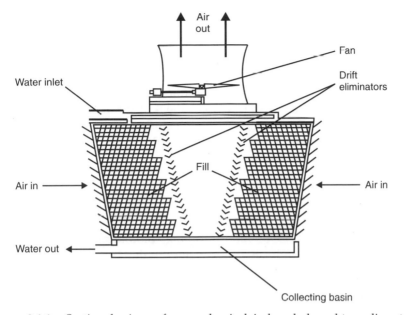

Figure 3.1.1. Sectional view of a mechanical–induced–draught cooling tower (Reprinted from Udaiyan and Manian *International Biodeterioration and Biodegradation* 1991, 351–372; with permission from Elsevier Science).

vary in length from metres to hundreds of kilometres! Thus it is almost impossible to be precise on where problem areas are likely to occur. The possibilities of corrosion in anaerobic muds and pipes carrying non-sterile liquids are frequently ignored. For example, in the oil production industry where secondary recovery systems require the injection of deoxygenated water down steel pipes into the oil bearing strata; many problems have developed with the growth of sulphate reducing bacteria (SRB) in the offshore section of the industry (Gaylarde 1989). NACE (National Association of Corrosion Engineers) list the following areas which can harbour SRB in the field: at stagnant points in flowlines, beneath scale deposits or debris in low velocity flowlines, in wash and water storage tanks, under sludge or in mud at the bottom of pits, on sand and gravel filters, at the oil-water interface in oil storage tanks and heater-treaters, and in the backfill on the outside of buried pipelines. All areas where the diffusion of oxygen is restricted. Hamilton (1985) describes the corrosion of steel drilling platform legs mediated by SRB biofilms releasing hydrogen sulphide, which attacked the metal. Attachment of bacteria such as the sulphate reducer, *Desulphovibrio vulgaris*, and the marine bacterium, *Vibrio alginolyticus*, to a metal surface has been shown to result in rapid metal corrosion (Gaylarde and Johnson 1980; Gaylarde and Videla 1987).

FILTRATION SYSTEMS

Filtration is used to control the size of particles in a fluid stream. Only in specialized small scale systems will that size include the smaller living organisms such as bacteria. Filtration units may consist of coarse and fine screens, fibre bag or mixed media deep-filter beds, followed by finer sand, cartridge or more novel types (Edyvean 1990). If filtration systems become blocked then the resulting problems will include the shut down of systems for cleansing and repair, loss of efficiency in fluid flow and heat exchange, and corrosion problems in pipeworks.

DESALINATION

Abd El Aleem *et al.* (1998) report fouling, biofouling and biocorrosion problems in membrane processes for water desalination and reuse in Saudi Arabia. They encountered the following adverse effects:

1. Membrane flux decline, resulting from the formation of a film of low permeability on the membrane surface.
2. Membrane degradation, brought about by the acidic by-products concentrated at the membrane surface.
3. Increased salt passage, brought about when biofilm formation leads to the accumulation of dissolved salt ions at the membrane surface, thus enhancing concentration polarization which has the effect of increasing salt passage through the membrane.
4. Increasing differential pressure and feed pressure, the danger here is damage to the membrane element as a result of the increased operating pressure brought about as a result of biofilm resistance and lower permeability.
5. An increased energy requirement related to the higher pressures needed to overcome the biofilm resistance and the flux decline.

METAL WORKING FLUIDS

Metal working fluids are used in industry to facilitate machining processes which involve either physical removal of metal from, or physical deformation of the material being machined (Prince and Morton 1988). Metal working fluids are needed to cool the workplace, to prolong tool life, to improve surface finish, to remove swarf, to reduce frictional heat between the tool and the chip, and to reduce power consumption. Any reduction in the efficiency of these functions constitutes a problem.

Metal working fluids are classified according to their chemical composition and fall into four major groups: neat oils, oil-in-water emulsions, 'semi-synthetic' oil in water emulsions and chemical solutions (American Society for Testing and Materials 1978). Oil in water emulsions are the most widely used industrial cutting fluids. Neat oils are not usually considered to be susceptible to biodeterioration since free water is seldom present (Seal and Morton 1986). The growth of microorganisms in oil-in-water emulsions can result in emulsion instability, lowering of pH, production of foul odours, the formation of other stable emulsions, and increased corrosive activity. The growth of microorganisms in 'semi synthetic fluids and chemical solutions' result in lowering of pH, production of fungal biomass and the production of foul odours. This microbial invasion can be readily appreciated when one considers that commonly employed additives in modern metal working fluids can be biodegraded by microorganisms. Additives include sodium or potassium sulphonates and carboxylic acids which are used as emulsifiers; esters of dimer acids or stearates which are used as extreme pressure additives; fatty oils, such as tall oil, rapeseed oil or lard which have been used as lubricity agents; and amine borate and triethanolamine which are employed as corrrosion inhibitors (Prince and Morton 1988). Synthetic fluids, which have evolved from chemical solutions are composed of three basic components namely, emulsifiers, lubricity agents (including polyglycols) and corrosion inhibitors. There are no reports on the bacterial degradation of whole synthetic fluids or their individual components (Buers *et al.* 1997). Synthetic metal working fluids have been reported to be susceptible to fungal contamination (Rossmore and Holtzman 1974; Prince and Morton 1988). It is not surprising therefore that bacterial biofilms and fungal biofilms are reported frequently at industrial sites, often associated with screen filters and swarf accumulations. Two major causes for concern arise from such occurrences. Firstly, why are the biocides which the formulations contain not effective? Secondly, do the infected formulations contain pathogenic microrganisms? The first of these questions will be dealt with later in this chapter. The answer to the second question is yes, pathogenic bacteria have frequently been isolated from in service fluids. Rossmore (1986) reports that pathogens have been isolated from metal working fluids often coexisting with non-pathogens. He also observes that workers may be exposed to fluids not only by contact, but also through inhalation of the fluids which may contain opportunistic pathogenic pseudomonads or *L. pneumophila* (Glick *et al.* 1978; Anon. 1991). The possibility of mycotoxin infection by any route is also a possibillity (Rossmore and Holtzman 1974).

Corrosion of Metals

Corrosion can be defined as the chemical reversion of a refined metal to its most stable energy state. In the extraction and refining process of a metal, energy is required to convert the ore to metal. During this process, chemical bonds are broken, oxygen, water and other anions removed and the pure metal arranged in an ordered lattice. Factors which encourage corrosion are those which will overcome the lattice bonding and encourage metal to leave the surface. The greatest of these is the presence of water or some other conductive electrolyte. Biologically influenced corrosion (BIC), microbial corrosion (MC) and microbial influenced or induced corrosion (MIC), are among the terms used to describe biological corrosion (Edyvean and Videla 1994). These authors report that a survey conducted in the USA in 1986 put the cost of corrosion at $167 billion with 10% of this attributable to the effects of biological attack.

Bacteria, algae and fungi may take part in the electrochemical reaction which is corrosion (Table 3.1.2). However, there are certain environments where the risk of MIC is greater than average (Gaylarde 1989). It should also be mentioned that protozoa, which are intimately associated with bacteria and fungi may influence fouling in open water systems, whilst macroalgae and invertebrates can be involved in fouling of offshore structures in the marine environment (Relini and Relini 1994). Biocorrosion is a major problem affecting most engineering alloys (de Mele and Videla 1994). The extent and nature of the corrosion sustained depends upon the metal and the ecology of the biofilm (Tiller 1990). The detailed mechanisms of MIC are beyond the scope of this chapter, however, they are reviewed (together with typical corrosion problems) in Gaylarde (1989), Edyvean and Videla (1994), de Mele and Videla (1994), Tiller (1990), Edyvean (1990) Lakatos (1990), Videla (1990), Videla and Characklis (1992) and Surman and Morton (1996). The mechanisms by which biofilms contribute to corrosion are influenced by the availability of oxygen in the environment. Under aerobic conditions, localized biofilm deposits can cause the formation of anodic and cathodic areas on the surface of a metal. These areas become a series of differential chemical cells, each inducing the transfer of electrons with loss of cations. Under aerobic conditions the utilization of the hydrocarbons of diesel fuel and aviation kerosene by *Hormoconis resinae*, results in the production of organic acids which are corrosive to metals (Hendey 1964; Parberry 1971; Hedrick 1970). Sulphur oxidizing bacteria produce sulphuric acid in quantities sufficient to bring about the corrosion of metals (Engvall 1986). Bacteria and fungi may accelerate corrosion indirectly by utilizing corrosion inhibitors (Prince and Morton 1988). Under anaerobic conditions sulphate reducing bacteria are the major cause of corrosion in low oxygen or oxygen-free environments. Gaylarde (1989) lists the various mechanisms by which sulphate reducing bacteria induce metal dissolution. They include:

1. Cathodic depolarization brought about by the bacterial enzyme hydrogenase,

Table 3.1.2. Microorganisms which may be involved in metal corrosion

Type of organism	Environment growth requirements	Main industrial problem areas
Green algae	Light	Cooling towers
Blue-green bacteria	Oxygen	Heat exchangers Metal structures in ponds and rivers
Most fungi and aerobic bacteria	Organic carbon. Oxygen	As above, plus metal structures in aerated soils
Pseudomonas	Oxygen Organic carbon May use complex or unusual carbon sources	As above, plus factory workshops where metal-working fluids (cutting oils) are used
Sulphur-oxidizing bacteria (e.g. *Thiobacillus*)	Oxygen. Sulphur or reduced sulphur compounds	Waste disposal plants (Also corrode concrete)
Iron-oxidising bacteria (e.g. *Gallionella*)	Oxygen Ferrous ions	Metal structures in open waters Oilfield injection wells
The fungus (*Hormoconis resinae*)	Oxygen Organic carbon (can use hydrocarbons)	Aircraft and marine fuel tanks and pipelines
Sulphate-reducing bacteria	Absence of oxygen Organic carbon. Sulphate ions	Buried pipes Beneath fouling layers Within closed or lengthy piping systems Metal workshops using cutting oils
Other anaerobic bacteria (e.g. *Clostridium*)	Absence of oxygen Organic carbon	As above

Table provided Dr C. Gaylarde.

2. The production of corrosive iron sulphides, through reaction of ferrous metals with the hydrogen sulphide released during bacterial metabolism,
3. Sulphide induced stress corrosion cracking,
4. Hydrogen induced cracking or blistering,
5. Oxidation of biogenic hydrogen sulphide to corrosive elemental sulphur, this can occur when oxygen is available in the environment.

Figure 3.1.2 (a) shows a scanning electron micrograph of an SRB film on mild steel with EPS and corrosion products, (b) and is the cleaned surface showing corrosion beneath.

BIOFOULING OF CONCRETE IN CONTACT WITH WATER

Since the introduction of Portland cement almost 50 years ago, concrete has become one of the most widely used synthetic materials within the construction industry (Neville 1977). Concrete is composed of cement powder, water and aggregates of various sizes, such as sand or gravel (Hueck van der Plas 1968). The main constituents of cement powder are lime, silica, alumina and iron oxide (Table 3.1.3).

Upon curing, the cement paste hydrolyses to form hydrated calcium silicates (C–S–H gels) and Portlandite ($Ca(OH)_2$) (Dreux 1964; Dubois 1977; Neville 1977; Darimont 1993). The arrangement of these crystals allows channels to form within the concrete structure. The formation of these channels within the concrete permits the capillary action of water. This water may contain microorganisms; thus, as water permeates deeper into the concrete, fissures are formed allowing the deposition of organic material and the further ingress of microorganisms into the concrete (Lea 1970).

History of Microbially Induced Corrosion of Concrete

In 1945, Parker reported that concrete was being broken down by the metabolic activities of microorganisms. His study was concerned with the isolation and identification of microbes from areas of badly corroded concrete where members of the genus *Thiobacillus* were isolated. This and other subsequent studies (Parker 1945; Parker and Prisk 1953) showed that thiobacilli were the causative agents of microbially induced corrosion of concrete structures. This view has been substantiated in recent years by several others working in the same field (Milde *et al.* 1983; Karavaiko and Zherebyat'eva 1989; Islander *et al.* 1991; Mori *et al.* 1992)

Biogenic Sulphuric Acid Corrosion of Concrete

The decay of stone or concrete by microorganisms is dependent on the production of corrosive metabolites which can solubilize minerals (Lewis *et al.* 1987). Freshly prepared concrete has a pH between 12 and 13 compared to the highly acidic pH (2.5–0.5) of corroded concrete. The incidence and activity of the thiobacilli are described in Sand and Bock (1987), Milde *et al.* (1983). Sand and Bock (1991) reported on the findings of a research project

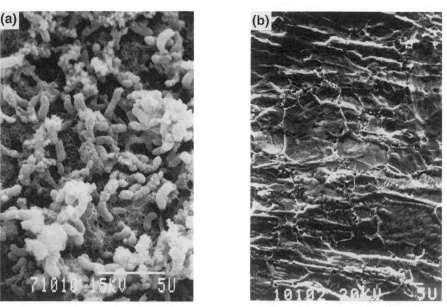

Figure 3.1.2. (a) A scanning electron micrograph of an SRB film on mild steel with EPS and corrosion products. (b) The cleansed surface showing corrosion beneath the film (by kind permission of Dr C Gaylarde).

undertaken in Hamburg to establish the mechanism of corrosion found in several pipelines constructed between 1967 and 1971, that thiobacilli were present in high numbers at the corrosion sites. The incidence of *Thiobacillus thiooxidans* was found to be high at sites which showed heavy corrosion. The bacterium was not detected at sites where corrosion was not detected. Thus it was considered appropriate to adopt this microorganism as an indicator of corrosion rather than the pH values resulting from bacterial sulphuric acid generation which are partially buffered by different concrete

Table 3.1.3. Main compounds of Portland cement

Name of compound	Oxide composition	Abbreviation
Tricalcium silicate	$3CaO.SiO_2$	C_3S
Dicalcium silicate	$2CaO.SiO_2$	C_2S
Tricalcium aluminate	$3CaO.Al_2O_3$	C_3A
Tetracalcium aluminoferrite	$3CaO.Al_2O_3Fe_2O_3$	C_4AF

After Neville (1977)

types. Volatile sulphur compounds, which are dissolved in the sewage, escape into the atmosphere within the sewer. In this atmosphere, hydrogen sulphide is autoxidised to molecular sulphur, which adheres to the concrete walls of the sewage pipes where it is further oxidised to sulphuric acid. This acid reacts with the calcitic binding material of the concrete causing its destruction. These workers also reported that two other sulphur compounds were detected on the sewer walls, namely sulphur dioxide and thiosulphate. Thiosulphate is a reaction product of sulphur dioxide with molecular sulphur and forms a good substrate for the growth of thiobacilli.

Sulphate Reducing Bacteria

Sulphate reducing bacteria are usually associated with metal corrosion rather than with the deterioration of concrete. However, sulphate reducing bacteria are widespread in nature and are active in locations made anaerobic by microbial digestion of organic materials. Figure 3.1.3 gives an overview of the sulphur cycle in a partially filled sewage pipe. It also shows the involvement of sulphate reducing bacteria in concrete decay as being mediated via gas produced by the proteolytic action of sulphate reducing bacteria living in anaerobic slimes or biofilms below the sewage level.

Biogenic Nitric Acid Corrosion of Concrete

Whilst thiobacilli are considered to be responsible for the degradation of concrete above ground, it is the two groups of nitrifying bacteria that are considered to play a major role in the degradation processes occurring below ground. They are responsible for the oxidation of ammonia via nitrous acid to nitric acid (Sand and Bock 1991). The first group include species of *Nitrosomonas*, *Nitrospira*, *Nitrosolobus*, *Nitrosovibrio* and *Nitroso-coccus*, which oxidize ammonia via hydroxylamine to nitrous acid. The second group which effect the oxidation of nitrous acid to nitric acid include members of the genus *Nitrobacter* and *Nitrococcus* (Prescott *et al.* 1996). The action of nitric acid upon the calcareous binding material of concrete results in the production of calcium nitrate, a soluble salt which is either lost from the concrete resulting in the formation of corrosion pits, or remains thus adding a salt to the pore water.

Biogenic Organic Acid Corrosion of Concrete

Mycelial development on a surface can be associated with the formation of mucilaginous sheaths. Sheath development in some genera is triggered by

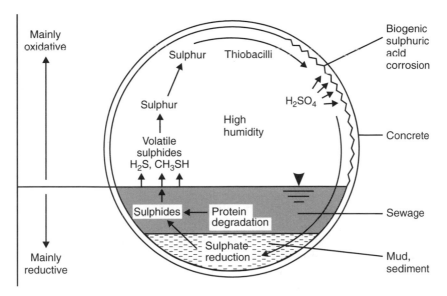

Figure 3.1.3. An overview of the sulphur cycle in a partially filled sewage pipe (Reprinted from Sand and Bock *International Biodeterioration and Biodegradation* 1991, 177; with permission from Elsevier Science).

physical contact with the substratum (Palmer *et al.* 1993; Jones 1994). Obvious similarities exist between the proposed functions of the fungal mucilages and the EPS of bacterial biofilms (Jones 1994; Jones 1995).

Little attention has been paid to the role of the fungi in the structure or nutrition of biofilms (Jones 1995). Fungal genera isolated from the surfaces of monuments, the facades of buildings and from degrading concrete include many genera which are representative of air and soil flora (Perichet 1987; Grant *et al.* 1989). Whilst these workers were not investigating industrial waters, their results point to the susceptibility of concrete to biodeterioration by heterotrophic microorganisms. McCormack *et al.* (1996) record the isolation of fungi from deteriorating concrete from a flooded cellar which formed corrosive biofilms on concrete in chemostat culture. Figure 3.1.4 shows a fungal biofilm on the surface of a concrete block.

THE RESISTANCE OF BIOFILMS TO BIOCIDES

Several laboratory studies have shown that microorganisms embedded within biofilms are protected from the lethal effects of biocides (Costerton 1984; Foley and Gilbert 1996). Whatever the mechanism, there is no doubt that in practical terms, the majority of biocides do not effectively penetrate biofilms (Gaylarde and Videla 1994). There are two mechanisms of bacterial resistance to biocides, namely intrinsic and acquired resistance (Russell 1996; Russell and Day 1996). It is considered that intrinsic resistance mechanisms may be more important in natural environments. Intrinsic resistance is a natural chromosomally controlled property or adaptation of an organism. Resistance to biocide treatments is increased in bacteria which are attached to surfaces and also to particulate matter within a system (Ridgway and Olsen 1982; Kutchta *et al.* 1985; King *et al.* 1988; Vess *et al.* 1993). A major role of the glycocalyx is that it constitutes a barrier affording the various constituents of the biofilm partial protection from antibacterial agents (Costerton *et al.* 1981; Cloete *et al.* 1989), and from the possible toxic effects of the substrate upon which a biofilm may form (e.g. copper pipes within distribution systems) (Keevil, *et al.* 1989a). Nichols (1994) reviews the mechanisms which may account for this protective property of the glycocalyx and suggests that the answer is not solely due to the physical impedance of the antimicrobial agent in question, but that there may be other factors such as adsorption or catalytic destruction of the agent by microbes on the biofilm surface. It is unclear whether a phenotypic response of the microbial population to surface growth also plays a role in this increased resistance (Jass and Lappin-Scott 1994). How this increased resistance is achieved is still unclear, and the answer may well prove to be a combination of several

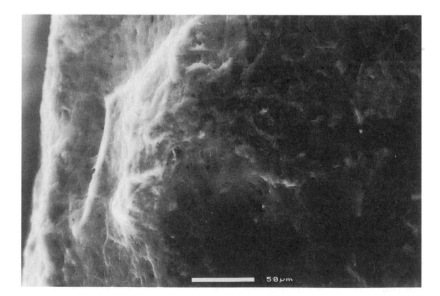

Figure 3.1.4.   A fungal biofilm growing on the surface of a concrete block (by kind permission of Miss K McCormack).

factors. Bacterial response to biocides is determined by the nature of the biocide and the type of organism involved. Other factors affecting the activity of a biocide include temperature, contact time, environmental pH and the presence of organic matter (Russell 1992). Nutrient limitation (Gilbert 1988) and reduced growth rates (Brown and Williams 1985) may also alter the sensitivity of bacteria to biocides. Cells on surfaces have different growth rates and nutritional requirements than planktonic cells of the same species and a number of workers have shown that biocide sensitivity can be altered up to 1000 fold by changes in nutrients and growth rates (Wright and Gilbert 1987). This increased resistance of biofilm microorganisms following treatment has also been observed in several studies.

1. The effect of a commercially available biocide on both mutant and 'wild type' *L. pneumophila* was studied in a model system (Keevil *et al.* 1987, 1989b) which contained both biofilm and planktonic associated legionellae and associated heterotrophs. Increased resistance of both the legionellae and associated heterotrophs was noted (Surman 1994; Surman *et al.* 1995; Surman *et al.* 1997). The extent of kill recorded for planktonic microorganisms after the first hour was significantly greater than that recorded with the sessile microorganisms. When the model systems were dosed with a second addition of biocide it was found that regrowth of the microorganisms occurred significantly faster in the sessile phase than in the planktonic phase.

2. A natural isolate of *Pseudomonas fluorescens* taken from the surface of the pool lining material of a domestic swimming pool was studied (Wilkinson 1994; Wilkinson *et al.* 1996). The attachment of *Ps. fluorescens* to PVC surfaces was rapid and irreversible, and clearly increased the organisms' resistance to the biocide, a polydisperse mixture of hexamethylene bisbiguanides (PHMB), by 2–4 times than exhibited by the planktonic population. The importance of growth rate may be of significance here since both the planktonic and sessile populations appeared to be in stationary phase but presumably the sessile organisms were growing much more slowly. These results are consistent with the views of Moyer and Morita (1989) who suggested that in oligotrophic, marine environments the cells exist and survive as slowly growing populations.

3. In a further study planktonic growth and biofilms of a species of *Pseudomonas* and *Ps. fluorescens*, were used to assess the biocidal efficacy of glutaraldehyde, formaldehyde, a mixture of isothiazolinones, ozone and sodium hypochlorite. Although all the biocides tested were effective in killing planktonic cells within the

range of concentrations used, their effect was markedly diminished when sessile populations were treated under the same experimental conditions and for the same contact times (Gaylarde *et al.* 1996).

The increased resistance of microrganisms in biofilms to biocides is now the focus of a great deal of research activity (Morton *et al.* 1998). A greater understanding of the mechanisms of resistance could play an important role in combating biofouling in industrial water systems.

REFERENCES

Abd El Aleem, FA, Al-Sugair, KA and Alamhad, MI (1998) Biofouling problems in membrane processes for water desalination and reuse in Saudi Arabia. *International Biodeterioration and Biodegradation* **41**:19–23.

American Society for Testing and Materials (1978) Standard classification of metal working fluids and related materials. ANSI/ASTM Designation D 2881-73 (Reapproved 1978). In Annual Book of ASTM Standards, Part 24, 750–751, ASTM, Philadelphia, PA.

Anon. (1991) Minimising the risk of legionnaires' disease. CIBSE Technical Memoranda 13.

Anwar, H, Van Biesen T, Dasgupta, MK, Lam, K and Costerton, JW (1989) Interaction of biofilm bacteria with antibiotics in a novel *in vitro* chemostat system. *Antimicrobial Agents and Chemotherapy* **33**:1824–1826.

Anwar, N and Strap, JL (1992) Changing characteristics of ageing biofilms. *International Biodeterioration and Biodegradation* **30**:177–186.

Beech, IB and Gaylarde, CC (1991) Microbial polysaccharides and corrosion. *International Biodeterioration* **27**:95–107.

Bentham, RH (1993) Environmental factors effecting the colonization of cooling towers by *Legionella* spp. in South Australia. *International Biodeterioration Biodegradation* **31**:55–63.

Brankevich, GJ, DeMele, MLF and Videla, HA (1990) Biofouling and corrosion in coastal power plant cooling water systems. *Marine Technology Society Journal* **24**:18–28.

Breiman, RF (1993) Modes of transmission in epidemic and non-epidemic legionella infection: Directions for further study. In: Legionella, current status and emerging prospectives American Society for Microbiology, The 4th International symposium on Legionella, Florida, 1992 (Eds. Barbaree, JM, Breiman, RF and Dufour, A), pp. 30–35.

Brown, MRW and Williams, P (1985) The influence of environment on envelope properties affecting survival of bacteria in infections. *Annual Reviews of Microbiology* **39**:527–556.

Bryers, JD and Characklis, WG (1982) Processes governing primary biofilm formation. *Biotechnology and Bioengineering* **24**:2451–2476.

Buers, KLM, Prince, ELM and Knowles, CJ (1997) The ability of selected bacterial isolates to utilise components of synthetic metal working fluids as sole source of carbon and nitrogen for growth. *Biotechnology Letters* **19**:791–794.

Characklis, WG and Cooksey, KE (1983) Biofilms and microbial fouling. *Advances in Applied Microbiology* **29**:93–138.

Cloete, TE, Smith, F and Steyn, PL (1989) The use of planktonic bacterial populations in open and closed recirculating water cooling systems for the evaluation of biocides. *International Biodeterioration* **25**:115–122.

Cofone, L, Walker, JD and Cooney, JJ (1973) Utilisation of hydrocarbons by *Cladosporium resinae. Journal of General Microbiology* **76**:243–246.

Colbourne, JS, Dennis, PJ, Trew RM, Berry C and Vesey G (1988) Legionella and public water supplies. *Water Science Technology* **20**:5–10.

Costerton, JW (1984) The formation of biocide-resistant biofilms in industrial, natural and medical systems. *Developments in Industrial Microbiology* **25**:363–372.

Costerton, JW, Cheng, K-J, Geesey, GG, Ladd, TI, Nickel JG, Dasgupta, M and Marrie, TJ (1987) Bacterial biofilms in nature and disease. *Annual Reviews of Microbiology* **41**:435–464.

Costerton, JW, Irvin, RT and Cheng K-J (1981) The bacterial glycocalyx in nature and disease. *Annual Reviews of Microbiology* **35**:299–324.

Costerton, JW, Nickel, JG and Ladd, TI (1986) Suitable methods for the comparative study of free living and surface associated bacterial populations. In: Bacteria in Nature, Vol. 2 (Eds. Poindexter, JS and Leadbetter, ER), Plenum Press. New York.

Darimont, A (1993) Concrete pathology—secondary precipitations. *Microscopy Research and Technique* **25**:179–180.

de Mele, MFL and Videla, HA (1994) Biological corrosion: A present state of art. In: Recent Advances in Biodeterioration and Biodegradation, Volume 11 (Eds. Garg, KL, Garg, N and Mukerji, KG), Naya Prokash, Calcutta, pp. 133-144.

Dempsey, MJ (1981a) Colonisation of antifouling paints in marine bacteria. *Botanica Marina* **24**:185–191.

Dempsey, MJ (1981b) Marine bacterial fouling: a scanning electron microscope study. *Marine Biology* **61**:305–315.

Dreux, G (1964) Connaissance du béton Société de diffusion des techniques du batement et des traveux publics, Lahure, Paris.

Dubois, J (1977) Le béton face aux agressions chimiques. *Chantiers Magazine* **80**:85–89.

Eastwood, IM (1994) Problems associated with biocides and biofilms. In: Bacterial Biofilms and their Control in Medicine and Industry (Eds. Wimpenny, J, Nichols, WW, Stickler, D and Lappin-Scott, H), British Biofilms Club Inaugural Meeting held at Gregynog Hall, Wales, Bioline, Cardiff, pp. 169–172.

Eaton, RA (1994) Bacterial decay of ACQ—treated wood in a water cooling tower. *International Biodeterioration and Biodegradation* **33**:197–208.

Eaton, RA (1976) Cooling tower fungi. In: Recent Advances in Aquatic Mycology (Ed. Jones, EBG), Elek Science, London, pp. 359–387.

Edyvean, RGJ (1990) Fouling and corrosion in water filtration and transportation systems. In: Micobiology in Civil Engineering (Ed. Howsman, P), E & FN Spon, London, pp. 62–74.

Edyvean, RGJ and Videla, HA (1994) Biological corrosion. In: Recent Advances in Biodeterioration and Biodegradation, Volume 11 (Eds. Garg, KL, Garg, N and Mukerji, KG), Naya Prokash, Calcutta, pp. 81–116.

Elsmore, R and Dorman, TW (1988) The development of a fouling monitor for the in-line measurement of biofouling. In: Biofilms (Eds. Morton, LHG and Chamberlain, AHL), Biodeterioration Society, Egham, pp. 14 25.

Engvall, A (1986) Biodeterioration of mineral materials. In: Biotechnology, 8 (Ed. Schonborn W), VCH Verlagsgesellschaft, Weinheim, pp. 607–626.

Exner, M, Tuschewitski, GJ and Scharnagel, J (1987) Influence of biofilms by chemical disinfection and mechanical cleaning. *Zentrablatt für Bakteriologie und Mikrobiologie und Hygiene Series B* **183**:549–563.

Ferris, FG, Schultze, S, Witten, TC, Fyfe, WS and Beveridge, TJ (1987) Metal interactions with microbial biofilms in acidic and neutral pH environments. *Applied and Environmental Microbiology* **55**:1249–1257.

Foley, I and Gilbert, P (1996) Antibiotic resistance of biofilms. *Biofouling* **10**:331–346.

Gaylarde, CC (1989) Microbial corrosion of metals. *Environmental Engineering* **2**:30–32.

Gaylarde, CC and Videla, HA (1994) The control of corrosive biofilms by biocides. *Corrosion Reviews* **2**:85–94.

Gaylarde, CC and Beech, IB (1989) Bacterial polysaccharides and corrosion. In: Biocorrosion, proceedings of joint meeting between the Biodeterioration Society and the French Microbial Corrosion Group (Eds. Gaylarde, CC and Morton, LHG), pp. 85–98.

Gaylarde, CC and Johnson, JM (1980) The Importance of microbial adhesion in anaerobic corrosion of mild steel. In: Microbial Adhesion to Surfaces (Eds. Berkeley, RCW, Lynch, JM, Melling, J , Rutter, PR and Vincent, B), Ellis Horwood, Chichester, pp. 511–551.

Gaylarde, CC and Videla, HA (1987) Localised corrosion induced by a marine vibrio. *International Biodeterioration* **23**:91–104.

Gaylarde, CC (1989) Microbial corrosion of metals. *Environmental Engineering* **2**:30–32.

Gaylarde, CC, Bento, FM, Viera, MR, Guiamet, PS, Gomez de Saravia, SG and Videla, HA (1996) Biocideal treatment of corrosive biofilms. Proc. 2nd Latin American Region Corrosion Congress, September 1996, NACE. Paper nos: LA96189.

Gilbert, P (1988) Microbial resistance to preservative systems. In: Microbial Assurance in Pharmaceuticals, Cosmetics and Toiletries (Eds. Bloomfield, SF, Baird, R Leak, RE and Leech, R), Ellis Horwood, Chichester, pp. 171–194.

Glick, TH, Greg. MB, Berman, B, Mallison, G, Rhodes Jr. WW and Kassanoff, I (1978) Pontiac fever an epidemic of unknown etiology in a health department: 1 Clinical and epidemiological aspects. *American Journal of Epidemiology* **107**:149–160.

Grant, C, Hunter, CA, Flannigan, B and Bravery. AF (1989) The moisture requirements of moulds isolated from domestic dwellings. *International Biodetererioration* **25**:259–284.

Grimes, DJ (1991) Ecology of estuarine bacteria capable of causing human disease. *Estuaries* **14**:334–360.

Gristina, AG, Salem, W and Costerton, JW (1985) Bacterial adherence to biomaterials and tissues. *Journal of Bone and Joint Surgery* **67A**:264–273.

Hamilton, WA (1985) Sulphate reducing bacteria and anaerobic corrosion. *Annual Reviews on Microbiology* **39**:195–217.

Heaton, PE, Callow, ME, Butler, GM and Milne, H (1991) Control of mold growth by antifungal paints. *International Journal of Biodeterioration and Biodegradation* **27**:163–173.

Heitz, E, Flemming, H-C and Sand, W (1996) Introduction. In: Microbially induced corrosion of materials (Eds. Heitz, E, Flemming, H-C and Sand), W, Springer, p. 1.

Hedrick, HG (1970) Microbial corrosion of aluminium. *Materials Protection* **9**:27–31.

Hendey, NI (1964) Some observations on *Cladosporium resinae* as a fuel contaminant and its possible role in the corrosion of aluminium alloy fuel tanks. *Transactions of the British Mycological Society* **47**:467–475.

Herbert, BN (1994) Biofilms and Pipelines. In: Bacterial Biofilms and their Control in Medicine and Industry (Eds. Wimpenny, J, Nichols, WW, Stickler, D and Lappin-Scott, H), British Biofilms Club Inaugural Meeting held at Gregynog Hall, Wales, Bioline, Cardiff, pp. 117–120.

Hill, EC (1975) Biodeterioration of petroleum products. In: Microbial aspect of the deterioration of materials (Eds. Gilbert, RJ and Lovelock DW), Academic Press, London, pp. 127–135.

Holah, JT, Bloomfield, SF, Walker, AJ and Spenceley, H (1994) Control of biofilms in the food industry. In: Bacterial Biofilms and their Control in Medicine and Industry (Eds. Wimpenny, J, Nichols, WW, Stickler, D and Lappin-Scott, H), British Biofilms Club Inaugural Meeting held at Gregynog Hall, Wales, Bioline, Cardiff, pp. 163–168.

Hueck van der Plas, EH (1968) The microbial deterioration of porous building materials. International Biodeterioration Bulletin 4:11–28.

Huek, HJ (1965) The biodeterioration of materials as a part of hylobiology. Material und Organismen 1:5–34.

Islander, RL, Devinny, JS, Mansfield, F, Postyn, A and Shih, H (1991) Microbial ecology of corrosion in sewers. Journal of Environmental Engineering 117:751–771.

Jass, J and Lapin-Scott, H (1994) Sensitivity testing for antimicrobial agents against biofilms. In: Bacterial Biofilms and their Control in Medicine and Industry (Eds. Wimpenny, J, Nichols, WW, Stickler, D and Lappin-Scott, H), British Biofilms Club Inaugural Meeting held at Gregynog Hall, Wales, Bioline, Cardiff, pp. 73–76.

Jones, EBG (1994) Fungal adhesion. Mycological Research 98:961–981.

Jones, MV (1995) Fungal biofilms: eradication of a common problem. In: The life and death of biofilms (Eds. Wimpenny, J, Handley, P, Gilbert, P and Lappin-Scott, H), Bioline, Cardiff, pp. 157–160.

Karavaiko, GI and Zherebyat'eva, TV (1989) Bacterial decay of concrete. Doklady Akademmi Nauk SSSR 306:355–359.

Keevil, CW, Walker JT, McEvoy J and Colbourne JS (1989a) Detection of biofilms associated with pitting corrosion of copper pipework in Scottish hospitals In: Biocorrosion, Proceedings of a joint meeting between the Biodeterioration Society and the French microbial corrosion grap (Eds. Gaylarde, CC and Morton, LHG), occasional publication No. 5, Biodeterioration Society, Eylam, pp. 99–117.

Keevil, CW, Bradshaw, DJ, Dowsett, AB and Feary, TW (1987) Microbial film formation: Dental plaque deposition on acrylic tiles using continuous culture techniques. Journal of Applied Bacteriology 62:129–138.

Keevil, CW, West, AA, Walker, JT, Lee, JV, Dennis, PJL and Colbourne, JS (1989b) Biofilms: Detection, implications and solutions. In: Watershed 89. The future for water quality in Europe, Volume II (Eds. Wheeler, D, Richardson, ML and Bridges, J), Pergamon Press, Oxford.

King, CH, Shotts, EB, Wooley, RE and Porter, KG (1988) Survival of coliforms and bacterial pathogens within protozoa during chlorination. Applied and Environmental Microbiology 54:3023–3033.

Kristinson, KG (1989) Adherence of staphylococci to intravascular catheters. Medical Microbiology 28:249–257.

Kutchta, JM, States, SJ, McGlaughlin, JE, Overmeyer, JH, Wadowsky, RM, McNamara, AM, Wolford, RS and Yee, RB (1985) Enhanced chlorine resistance of tap water-adapted Legionella pneumophila as compared with agar medium passaged strains. Applied and Environmental Microbiology 50:21–26.

Lakatos, G (1990) Study on biofouling forming in Industrial cooling water systems. In: Microbiology in Civil Engineering (Ed. Howsman. P), E & FN Spon, London, pp. 80–94.

Lea, FM (1970) The chemistry of cement and concrete. 3rd Edition. Edward Arnold, London, pp. 591–593.

LeChevallier, MW, Cawthon, CD and Lee, RG (1988a) Factors affecting the survival of bacteria in chlorinated water supplies. Applied and Environmental Microbiology 54:649–654.

LeChevallier, MW, Cawthon, CD and Lee, RG (1988b) Inactivation of biofilm bacteria. *Applied and Environmental Microbiology* **54**:2492–2499.

LeChevallier, MW, Cawthon, CD and Lee, RG (1988) Factors affecting the survival of bacteria in chlorinated water supplies. *Applied and Environmental Microbiology* **54**:649–654.

Lewis, FJ, May, E and Bravery, AF (1987) The role of heterotrophic bacteria and fungi in the decay of sandstone from ancient monuments. In: Biodeterioration of constructional materials (Ed. Morton, LHG), Biodeterioration Society, Egham, pp. 45–53.

Lieve, L, Shoukin VK and Mergenhagen SE (1968) Physiological, chemical and immunological properties of LPS released from *E. coli* by EDTA. *Journal of Biological Chemistry* **243**:6384–6391.

Lynch, JL and Edyvean, RGJ (1988) Biofouling in oilfield water systems: a review. *Biofouling* **1**:147–162.

Marsh, PD, Bradshaw, DJ, Watson, GK and Cummings, D (1993) Factors affecting the development and composition of defined mixed biofilms of oral bacteria. In: Bacterial Biofilms and their Control in Medicine and Industry (Eds. Wimpenny, J, Nichols, WW, Stickler, D and Lappin-Scott, H), British Biofilms Club Inaugural Meeting held at Gregynog Hall, Wales, Bioline, Cardiff, pp. 7–9.

McCormack, K, Morton, LHG, Benson, J, Osborne, BN and McCabe, RW (1996) A preliminary assessment of concrete biodeterioration by microorganisms. In: Biodegradation & Biodeterioration in Latin America (Eds Gaylarde, CC, de Sá, ELS and Gaylarde, PM), Porto Alegre, Mircen/UNEP/UNESCO/ICRO-FEPAGRO/UFRGS, pp. 68–70.

McCoy, WF, Bryers, JD, Robbins, J and Costerton, W (1981) Observations of fouling biofilm formation. *Canadian J. Microbiology* **27**:910–991.

McCoy, WF, Bryers, JD, Robbins, J and Costerton, W (1981) Observations of fouling biofilm formation. *Canadian Journal of Microbiology* **27**:910–991.

Milde, K, Sand, W, Wolff, W and Bock, E (1983) Thiobacilli of the corroded concrete walls of the Hamburg sewer system. *Journal of General Microbiology* **129**:1327–1333.

Mori, T, Nonaka, K, Tazaki, K and Hikosaka, Y (1992) Interactions of nutrients, moisture and pH on microbial corrosion of sewer pipes. *Water Research* **26**:29–37.

Morton, LHG, Greenway, DLA, Gaylarde, CC and Surman, SB. (1998) Considerations of some implications of the resistance of biofilms to biocides. *International Biodeterioration and Biodegradation* **41**:247–260.

Moyer, CL and Morita, RY (1989) Effect of growth rate and starvation-survival on the viability and stability of a psychrophilic marine bacterium. *Applied and Environmental Microbiology* **55**:1122–1127.

Neville, AM (1977) The Properties of Concrete. 2nd edition, Pitman Publishing Ltd., London.

Nichols, W (1994) Biofilm permeability to antibacterial agents. In: Bacterial Biofilms and their Control in Medicine and Industry (Eds. Wimpenny, J, Nichols, W, Stickler, D and Lappin-Scott), Bioline, Cardiff, pp. 141–150.

Palmer, RJ, Siebert, J and Hirsch, P (1993) Biofilms and organic acids in sandstone of a weathering building—production by bacterial and fungal isolates. *Microbial Ecology* **21**:251–266.

Parberry, DG (1971) Physical factors influencing growth of *Amorphotheca resinae* in culture. *International Biodeterioration Bulletin* **7**:5–9.

Parker, CD (1945) The corrosion of concrete 1. The isolation of a species of bacterium associated with the corrosion of concrete exposed to atmospheres containing hydrogen sulphide. *Australian Journal of Experimental Biology and Medical Science* **23**:81–90.

Parker, CD and Prisk, J (1953) The oxidation of inorganic compounds of sulphur by various sulphur bacteria. *Journal of General Microbiology* **8**:344–364.

Perichet, A (1987) Biodeterioration study of facade materials with hydraulic binders. In: Biodeterioration of constructional materials (Ed. Morton, LHG), Biodeterioration Society, Egham, pp. 103-112.

Peterson, PK and Quie, PG (1981) Bacterial surface components and the pathogenesis of infectious diseases. *Annual Review of Medicine* **32**:29–43.

Prescott, LM, Harley, JP and Klein, DA (1996) Micobiology 3rd Edition, WCB Publishers, Oxford, p. 465.

Prince, EL and Morton LHG (1988) Fungal biodeterioration of synthetic metal working fluids, In: Biofilms (Eds. Morton, LHG and Chamberlain, AHF), Biodeterioration Society, Egham, pp. 107–122.

Reasoner, DJ (1988) Drinking water microbiology research in the United States: an overview of the past decade. *Water Science Technology* **20**:101–107.

Reid, G, Busscher, HJ (1992) Importance of surface properties in bacterial adhesion to biomaterials, with particular reference to urinary tract. *International Biodeterioration and Biodegradation* **30**:105–122.

Relini, G and Relini, M (1994). Macrofouling on offshore structures in the Mediterranean sea. In: Recent Advances in Biodeterioration and Biodegradation, Volume 11 (Eds. Garg, KL, Garg, N and Mukerji, KG), Naya Prokash, Calcutta, pp. 307–326.

Ridgway, HF and Olson, BH (1982) Scanning electron microscope evidence for bacterial colonisation of a drinking-water distribution system. *Applied and Environmental Microbiology* **41**:274–287.

Rittenhouse, RC (1991) Industry weapons grow in biofouling battle. *Power Engineering* **95**:17–23.

Rossmore, HW (1986) Microbial degradation of water-based metal working fluids. In: Comprehensive Biotechnology (Ed. Moo-Young, M), Pergamon, Oxford, pp. 249–269.

Rossmore, HW and Holtzman, GH (1974) Growth of fungi in cutting fluids. *Developments in Industrial Microbiology* **15**:273–280.

Russell, AD and Day, MJ (1996) Antibiotic and biocide resistance of bacteria. *Microbios* **85**:45–65.

Russell, AD (1996) Mechanisms of bacterial resistance to biocides. *International Biodeterioration and Biodegradation* **36**:247–265.

Russell, AD (1992) Plasmids and Bacterial Resistance. In: Principles and Practice of Disinfection, Preservation and Sterilisation, 2nd Edition (Eds. Russell, AD, Hugo, WB and Ayliffe, GAJ), Blackwell Scientific Publications, Oxford, pp. 225–229.

Sand, W and Bock, E (1987) Simulation of biogenic sulphuric acid corrosion of concrete—Importance of hydrogen sulphide, thiosulphate and methylmercaptan. In: Biodeterioration of Constructional Materials (Ed. Morton, LHG), Biodeterioration Society. Egham, pp. 29–36.

Sand, W and Bock, E (1991) Biodeterioration of mineral materials by microorganisms—Biogenic sulphuric acid and nitric acid corrosion of concrete and natural stone. *Geomicrobiology Journal* **9**:129–138.

Savory, JG (1964) Breakdown of timber by ascomycetes and fungi imperfecti. *Annual Applied Biology* **41**:336–347.

Seal, KJ and Morton, LHG (1986) Chemical materials. In: Biotechnology. Vol. 8 (Eds. Rehm, HG and Reed, G), VCH Verlagsgesellschaft, Weinheim, pp. 595–599.

Shariff, N and Hassan, RS (1985) Engineering and nutritional parameters affecting biofilm development. *Effluent and Water Treatment Journal* **25**:423–425.

Shaw, JC, Bramhill, B, Wardlaw, NC and Costerton, JW (1985) Bacterial fouling in a model core system. *Applied and Environmental Microbiology* **49**:693–701.

Singh, AP, Hedley, ME, Page, DR, Han CS and Atisongkroh, K (1992) Microbial degradation of CCA—treated cooling tower timbers. *International Association of Wood Anatomists Bulletin* **13**:215–231.

Surman, SB, Morton, LHG and Keevil, CW (1995) The use of a biofilm generator in the evaluation of a biocide for use in water treatment. In: Biodeterioration and Biodegradation 9; Proceedings of the 9th International and Biodeterioration and Biodegradation Symposium (Eds. Bousher, A, Chandra, M and Edyvean, R), I. Chem. E., Rugby, U.K., pp. 7–16.

Surman, SB (1994) The integration of an avirulent *L. pneumophila* into aquatic biofilms. Ph.D. Thesis, University of Central Lancashire, Preston, U.K.

Surman, SB and Morton, LHG (1996) Bi-ofilms: an overview. *PHLS Microbiology Digest* **13**:33–38.

Surman, SB, Goddard, D, Morton, LHG and Keevil, CW (1997) The use of an avirulent *Legionella pneumophila* for the evaluation of a biocide. In: Bioflims: community interaction and control (Eds. Wimpenny, J, Handley, P, Gilbert, P, Lappin-Scott, H and Jones, M), Bioline, Cardiff, pp. 269–278.

Tiller, AK (1990) Biocorrosion in civil engineering. In: Microbiology in Civil Engineering (Ed. Howsman, P), E &FN Spon, London, pp. 24–38.

Tobin, JOH, Swann, RA and Bartlett, CLR (1981) Isolation of *Legionella pneumophila* from water systems: methods and preliminary results. *British Medical Journal* **282**:515–517.

Trulear, MG and Characklis WG (1982) Dynamics of biofilm processes. *Journal of Water Pollution Control Federation* **54**:1288–1301.

Udaiyan, K and Manian, S (1991) Fungi colonising the wood in the cooling tower water systems at the Madras Fertilizer Company, Madras India. *International Biodeterioration and Biodegradation* **27**:351–372.

Väisänen, GM, Nurmiaho-Lassila, E, Marmo, SA and Salkinoja-Salonen, MS (1994) Structure and composition of biological slimes on paper and board machines. *Applied and Environmental Microbiology* **60**:641–653.

van der Wende, E, Characklis, WG and Smith, DB (1989) Biofilms and drinking water quality. *Water Research* **23**:1313–1322.

Vess, RW, Anderson, RL, Carr, JH, Bond, WW and Favero, MS (1993) The colonisation of solid PVC surfaces and the acquisition of resistance to germicides by water microorganisms. *Journal of Applied Bacteriology* **74**:215–221.

Videla, HA (1990) Microbially induced corrosion: an updated overview. In: Biodeterioration and Biodegradation 8, Proc. of the 8th International Biodeterioration and Biodegradation Symposium, Windsor Ontario (Ed. Rossmore HW), Elsevier Science Publishers. London, pp. 63–89.

Videla HA and Characklis, WG (1992) Biofouling and microbially influenced corrosion. *International Biodeterioration and Biodegradation* **29**:195–212.

Walker, JT and Keevil, CW (1994) Study of microbial biofilms using light microscope techniques. *International Biodeterioration and Biodegradation* **34**:223–236.

Wilkinson, DT (1994) Studies of Aquatic Microorganisms in Domestic Swimming Pools. Ph.D. Thesis, University of Central Lancashire, Preston, U.K.

Wilkinson, DT, Greenway, DLA, Eastwood, I and Morton, LHG (1996) Isolation and characterisation of microorganisms from two private swimming pools. *Pollution Control Bulletin* **5**:8–9.

Wright, NE and Gilbert, P (1987b) Influence of specific growth rate and nutrient limitation upon the sensitivity of *Escherichia coli* to chlorhexidine diacetate. *Journal of Applied Bacteriology* **62**:309–314.

Zobell, CE and Allen, EC (1935) The significance of marine bacteria in the fouling of submerged surfaces. *Journal of Bacteriology* **25**:230–251.

3.2 Detection of Biofilms in Industrial Waters and Processes

STEVEN L. PERCIVAL
Senior Lecturer in Medical Microbiology,
Department of Biological Sciences, The University of
Central Lancashire, Preston, PR1 2HE, UK

INTRODUCTION

Biofouling or biofilm development on engineered materials can lead to substantial effects on the performance of industrial processes. The undesirable accumulation of biological/chemical growth can ultimately lead to a reduction in the efficiency of equipment and thus a reduction of its life expectancy. Problems often encountered by engineers within industrial systems include corrosion, deterioration of materials, loss of heat transfer in heat exchangers, increased pumping costs, odour, blockage of filters and spoilage of the product (King 1996).

As fouling is classed as the formation of deposits on a surface, there is the possibility of a number of definitions (Characklis 1985). Whilst these have been strictly classified, it must be stressed that within most operating systems, the type of fouling may occur simultaneously. The different types of biofouling associated with industrial applications include (Characklis 1985):

1. Biological fouling—This can either be due to the formation and build up of macroorganisms or microorganisms.
2. Chemical reaction fouling—These are deposits formed by chemical reactions which do not involve the substratum as a reactant.
3. Corrosion fouling—This forms as a result of the substratum reacting with the bathing fluid to which it is exposed.
4. Freezing fouling—This is due to the solidification of a liquid.
5. Particulate fouling—This involves the build up of finely divided solids on equipment.
6. Precipitation fouling or scaling—This process is formed by the precipitation of substances on surfaces.
7. Sedimentation fouling—The gravity settlement of material on a surface.

Industrial Biofouling. Edited by J. Walker, S. Surman, J. Jass
©2000 John Wiley & Sons Ltd

The cost of both fouling and corrosion within industrial and process waters is colossal, with corrosion estimated at 3–4% of the gross national product of a country (Payer *et al.* 1978). The effects associated with microbial growth in industry has been estimated at 0.5% GNP for the U.K. (Pritchard 1981).

THE NEED FOR BIOFILM DETECTION

A biofouling detection system is necessary within an industrial setting in order to minimize any effects it has on performance of a plant. Information which can be gained about biofouling within industrial processes is important in order to prepare for any future problems, interventions and evaluation of biocide effectiveness. Therefore, a detection method is necessary to evaluate biofouling and biofilm development for future investment.

A general inspection of an industrial process initially involves a visual examination to make a preliminary assessment of biofouling and its associated effects. This provides the first approach to a monitoring programme. However, visual examination as a means of biofouling detection is generally confined to open conduits and cooling towers where the problem is clearly visible. Reliable data are necessary to fully evaluate biofouling and its potential problems. If adequate data were obtained on full-scale equipment, then small-scale laboratory tests and detection methods would not be necessary. However, it is not possible to interpret operations data adequately to evaluate problems due to fluctuating parameters. Therefore, laboratory based investigations are necessary. This allows for carefully controlled parameters to be investigated.

Within industrial processes, prior to more automated monitoring systems, engineers relied upon periodic inspections of plants as a means of assessing the fouling/corrosion of an engineering process. This preliminary method of biofouling detection cannot be used as the sole detection system, as the actual visual detection of biofouling would suggest that the biofilm had progressed to an advanced stage with secondary problems. More sophisticated methods are therefore necessary, if biofilm development and its control are to be fully investigated. The use of a more complex monitoring system would prevent annual shutdowns and reduce financial losses, a situation often observed during visual inspections of industrial plants.

INDUSTRIAL APPLICATIONS WHICH REQUIRE BIOFILM DETECTION

Within the industrial context, biofouling is associated, and can be regarded as a problem, in areas ranging from cooling water systems (once-through and recirculating systems), potable and waste water treatment plants, storage tanks, filters, reverse osmosis membranes, ion exchangers and piping systems. Whilst microbial biofouling will be concentrated on within this chapter, biofouling can also be associated with other types of fouling including crystalline or precipitate fouling (scaling) and particulate fouling.

GENERAL APPROACH OF BIOFOULING DETECTION

The current methods used in the detection and ultimately control of biofouling are complex. Particularly since biofilm formation is influenced by physical, chemical and biological factors determined by the exposed substratum to a bathing fluid. A large number of industrial organizations are now relying on computerized modelling systems to predict biofouling problems, thus often avoiding the expensive laboratory based approach. Whatever the technique used to monitor biofouling within the industry, however, there is also a need to undertake conventional microbial enumeration as a means of assessing the extent of biofouling. This is usually done in the laboratory after sampling of the fluids or surfaces in question. The type of sampling method employed is critical to the quality of the data derived from the analysis. With respect to the industrial environment, different sampling techniques and guidance from specific industrial recommended practices should be obtained (Miller 1971; NACE 1990).

Whilst detection is a necessity, experimental investigations of fouling is a prerequisite to the monitoring process to minimize the effects on an industrial application. If reliable data can be obtained, appropriate control measures can then be implemented accordingly.

FIELD TESTS APPLIED TO A DETECTION SYSTEM

Measurements at the industrial process site are both beneficial and essential. A number of field methods and observations can provide information on biofilms and biofouling (Flemming 1991). These include:

- Visual inspection of surfaces.
- Slimy layer—detected by touching the accessible area with a finger or clean cloth.
- 'Prolytic field test'—this provides an indication of the biological nature of the biofilm.
- Microscopic methods using specific dyes for interpretation.

Fouling monitors simulate the process environment and can be used to evaluate the potential for fouling as well as the effectiveness of any treatment programme. These monitors, such as sidestream fouling monitors, are used extensively in power plants and are capable of quantitatively simulating fouling processes in condensers. Sidestream monitors are also useful for testing the effects of various design, operating and environmental variables on the biofouling process. However, *in situ* fouling monitors are needed for regulating the control progress in the condenser. More sensitive fouling monitors (sidestream and *in situ*) are needed to detect very thin biofilms (and their effects) and to permit the rapid feedback for fouling control treatments. Frequently, on site fouling test systems are only connected for several months. There may be slow processes which only manifest themselves in the fouling dynamics after longer periods of time. However, monitors installed *in situ* will supply long term data that can serve as calibrating models. Fouling monitors are capable of quantitatively simulating fouling processes as they occur in operating equipment.

LABORATORY TESTS APPLIED TO BIOFOULING DETECTION

Laboratory tests are generally more cost effective than field tests, when goals and objectives are defined (Flemming and Geesey 1991). The laboratory provides environments for conducting defined tests that may lead to useful model systems simulating the 'real world'. Laboratory tests, therefore, provide the starting point for further investigation. Information can be obtained on the specific substratum being used within an industrial process and its association with biofouling, fluid velocity, tube geometry and temperature profiles. The effects of water quality can also be evaluated. Laboratory tests can also evaluate a detection system and assess the appropriate control measures to be employed. These tests frequently identify the operating conditions which are best for a given treatment procedure. They often result in the development of more novel measurement techniques, useful for field tests and monitoring.

Flemming and Geesey (1991) suggests that once biofouling is detected using field tests, the actual identification of the type of biofouling is based on laboratory investigations. They proposed that this includes one or all of the following methods (these will be discussed in more detail later in this chapter):

- Microscopic inspection
- Cultivation of microorganisms

• Analysis of surface deposits for water content, organic matter (total organic carbon), protein, carbohydrate, pyrogens, fatty acids, (adenosine triphosphate (ATP) and the activity of enzymes.

Flemming (1991) has also suggested that biofilms can be characterized by their high water content (70–95%), high organic matter content (50–95%), high microbial cell numbers, high protein and total carbohydrate content and high ATP content.

DETECTION AND ENUMERATION OF MICROORGANISMS WITHIN PROCESSES

Whilst on-line monitoring systems are available for the detection of biofouling, it is often beneficial to undertake conventional microbial enumeration of both the bathing fluid (planktonic cells) and surface (sessile cells) within any process that may be susceptible to fouling. However, the commonly used practice of monitoring within industrial waters still involves the estimation of bacterial numbers and genera within the planktonic state rather than the sessile state. This results in a large underestimation of the true microbiological problem evident upon a surface (Bott and Miller 1983). Often within cooling water systems, planktonic total cell counts are performed to indicate the extent of microbial contamination. However, this will not determine changes within bacterial, fungi and algae numbers and genera. Groups of these organisms need to be monitored separately since they are known to cause changes within the industrial water chemistry and biology. A specific known case is of corrosion initiated and accelerated by sulphate reducing bacterial (SRB) levels (Videla 1996).

It is well documented that cell numbers present within the planktonic phase do not correlate with the extent, location and viability of biofilms (Flemming and Geesey 1991). Flemming and Geesey (1991) have also suggested that microorganisms in biofilms are not released proportionately into the bathing fluid and that suspended cell numbers do not reflect an assessment of the sessile levels of the microorganisms. It is therefore, becoming a routine operation to assess the sessile microbial content associated with industrial process waters. The assessment of the microbial population has a number of distinct advantages (Videla 1996). Firstly, biofilms are more representative of the microbial population of the industrial process, which is particularly relevant to corrosion. Sessile bacterial numbers have been shown to be 3–4 logarithm units higher when compared to their planktonic counterparts. Secondly, the microbial species that predominate in the bulk phase may not be the same as those present within the biofilm. Finally, sessile microbial populations differ substantially in their susceptibility to antimicrobial agents when compared to their planktonic counterparts.

Formation of biofilms on wetted surfaces is recognized as an important factor in the control of corrosion, fouling and microbiological activity. The deleterious effects of biofilm growth within the industrial setting includes the following:

1. Energy losses due to reduced heat transfer in heat exchanges—biofilms have been estimated to cause a 30% reduction in 30 days.
2. Reduction in plant efficiency, reduction in water flow rates, insulation of heat-exchange surfaces, blockage of valves, filter traps, dead-legs etc.
3. Microbiologically induced corrosion activity, formed beneath biofilms due to the activities of SRBs can result in premature failure of pipework—it has been estimated that about 10% of corrosion is caused by microbiological activity.
4. An environment where potentially harmful bacteria, particularly *Legionella* species can grow. It is well documented that the biofilm provides both nutrients and protection for the constituent microbes (Anon. 1991).

Biofilm monitoring and control ensures compliance with the health and safety legislation. The Health and Safety Executives' guidance document HS(G)70 has emphasized the importance of effective biofilm control in reducing the risk of *Legionella* contamination. Within the document, the following quote illustrates the importance of biofilm monitoring and detection:

> "The formation of a biofilm within a water system is thought to play an important role in harbouring and providing conditions in which *Legionella* can grow. Incorporation of *Legionella* in biofilms can protect the organism from concentrations of biocides that would otherwise kill or inhibit those organisms freely suspended in water."

SAMPLING DEVICES AS PART OF A BIOFOULING DETECTION SYSTEM

Due to the detrimental effects of biofouling, there is need for *in situ* sampling devices to enable further analysis of the biofilm. This seems to be a prerequisite for testing the efficiency and evaluating the potential of biocides on reducing the biofouling process (Green *et al.* 1987). These 'sampling devices' are generally divided into 4 separate classes.

1. Monitoring devices directly implanted into the normal process, often referred to as insertion equipment.

2. Side-stream devices attached onto the engineered system, often in parallel to the main system under study (Videla 1994).
3. Recirculating rig systems.
4. Tubular geometry biofilm monitoring devices (Cloete *et al.* 1992).

Each of these systems will be considered in turn.

Direct Monitoring Devices (Probes)

Direct monitoring devices or probes are inserted directly into the industrial system to be tested. An example of their use is that associated with water injection lines for oil production. These systems are generally compatible to the normal industrial process material. Examples of the most commonly encountered sampling devices within industrial plants include the Caproco (Figure 3.2.1), Rohrbach-Cosasco and the Petrolite Bioprobe™ (Figure 3.2.2).

These systems are used for studying the effects of biocides on biofilms. However, difficulties may arise due to the variability of operating parameters that are very difficult to control. A disadvantage of this system is that the access fittings are small, therefore the size of the studs and the number of studs is limited (King 1996). Also, removing the fittings require both skill and dexterity. The advantages of this system are that biofilms formed on the probes are more representative to those formed in the main industrial process. These will vary with the process conditions. Ideally with

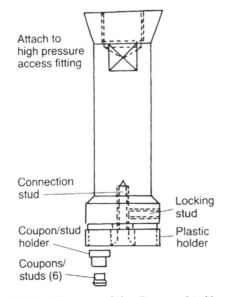

Figure 3.2.1. Diagram of the Caproco biofilm probe.

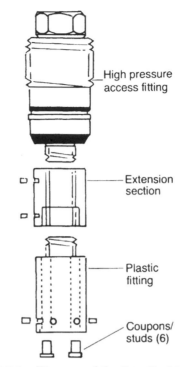

High pressure
access fitting

Extension
section

Plastic
fitting

Coupons/
studs (6)

Figure 3.2.2. Diagram of the Petrolite bioprobe.

this system, the device should be fitted directly into the pipe wall and it must be compatible with the existing fittings to avoid problems.

Side-stream Devices

These devices are generally installed parallel to the main engineered system. This enables the device to be exposed to identical operating conditions to that of the system under investigation. Therefore, the use of the actual process water can ensure that the biofilm is comparable to those that are formed within the main pipework. A side-stream system that is commercially available and used extensively to study biofilm development in flowing systems is the Robbins device (see Figure 3.2.3, McCoy *et al.* 1981; Ruseska *et al.* 1982).

It has now been modified to a more user-friendly laboratory based model, made from perspex rather than brass. Generally, coupons are used to investigate biofouling on different surfaces. The use of side-stream devices to evaluate biofilms within cooling water systems has been well documented (Bott 1992).

The general disadvantage of this device is the very small sampling areas used when compared to the actual pipe surface. Due to its time-consuming operation and complex construction, the Robbins device has been a subject of much debate and discussion. Another principal disadvantage is that it cannot model the hydrodynamics or geometries of the main flow, resulting in a poor evaluation of these parameters and thus may not be a true representation of the 'real' environment.

There is an increase in the use of small portable rigs that are transported by the service company to the plant. These systems operate under low pressure and are used to measure the extent of biofilm development, corrosion, scaling and also enables the investigation of biocide efficacy.

Side-stream devices offer greater flexibility to the user than the directly implanted devices. In addition, they can be located away from the main stream and, subsequently, connected to flow loops for biocidal testing. Besides the Robbin's device, a number of other side-stream devices have been developed over recent years (Gilbert and Herbert 1987; Green *et al.* 1987; Sanders and Hamilton 1986). One such device is called the RENAprobe[Tm] (REusable, Non-conventional Appliance). The RENAprobe was generally developed to study both biofouling and biocorrosion and to simulate these effects within recirculating cooling water systems or oil filled injection lines. Usually the sampler consists of a Teflon[TM] holding rod, alternately drilled on opposite faces to hold eight round metal coupons (of 0.5 cm^2 area), four on each face. The sampler can be directly inserted into non-pressurized areas of the flow system or, generally, as a side-stream device placed in a conventional corrosion rack.

Another side-stream monitoring device developed is the Houseman biofilm monitor (Houseman, Slough, UK), however, its usage is not very widespread (Figure 3.2.4). This biofilm monitor can be installed in hot and

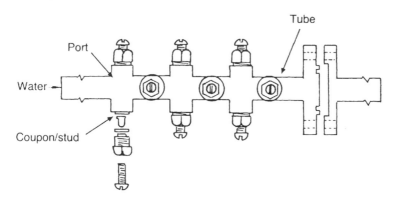

Figure 3.2.3. Diagram of the Robbins device used in the detection of biofilms.

cold water distribution systems, ideally as far as possible from the inlet, the storage tank or (for hot water systems) the calorifier. The unit can be located in concealed locations (such as plant rooms and riser conduits). The unit is enclosed in a lockable casing and consists of four biostuds which simulate the interior surface of the water system. These studs can be aseptically removed and the biofilm on their surface resuspended in sterile saline or double distilled water.

The Fospur biofilm monitor is also a side-stream device produced by Ashland U.K. Ltd. (Somercotes, Derbyshire). This system incorporates up to 4 bio-plates (total surface area of 35 cm^2 each), provided in a range of metals

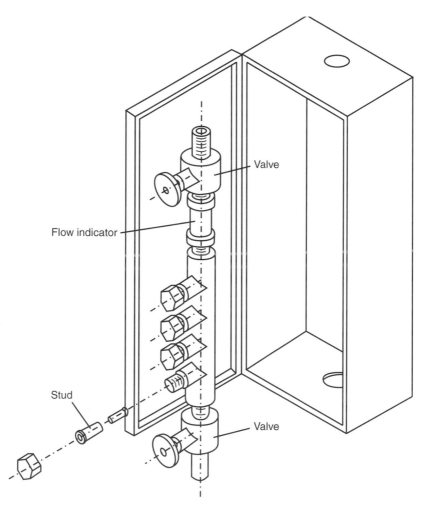

Figure 3.2.4. The Houseman biofilm monitor.

or plastics to suit individual system characteristics, on which the biofilm can form and be monitored.

In an industrial setting, there is a need for a simple, rapid, direct testing method which can be part of the routine quality control system. This can be achieved by use of the Bioprobe luminometer (Hughes Whitlock Limited, Gwent, U.K.) in conjunction with the biotube (Anon. 1997). The Bio-tube is a side-stream collecting device housing a removable arm onto which collection plates (coupons) are attached and the bioprobe is a sensitive luminometer which enables operatives to take measurements (Figures 3.2.5 and 3.2.6). Measuring coupons immediately without the need to scrape or disrupt the biofilm will enable rapid assessment of microbial contamination in process waters and pipe lines in one step. The advantage of this system is the reduction in expensive sample analysis and the provision of an early warning when there is a need for sanitation thus improving plant efficiency. Biofilm measurements in 'real time' allows for the efficacy testing of biocide treatment, leading to a reduction of pathogen proliferation in the biofilm. The benefits of the Bio-tube over previously used monitors are:

1. Each detection plate displays 30% of the surface of the internal circumference of the pipe.
2. The removal and replacement of collector plates is quick and easy, since they are simply held in place by two nylon screws.
3. Bio-tubes are manufactured to take one- or two- 10 × 35 mm collector plates, or two- or four- 50 × 35 mm plates.
4. It is fitted with 1 inch British Standard Pipe female thread at each end.

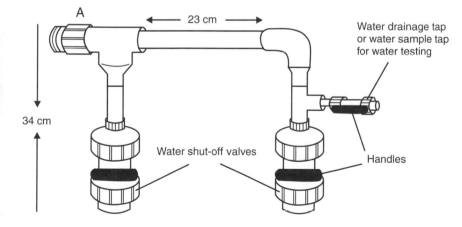

Shape and size can be altered if required

Figure 3.2.5. The biotube becomes a part of the system and therefore provides an insight to the presence of biofilm in the water system.

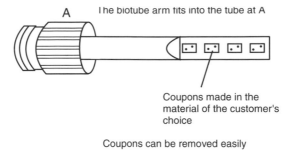

A   The biotube arm fits into the tube at A

Coupons made in the
material of the customer's
choice

Coupons can be removed easily

Protocol: • Remove coupons. Place on sample holder
• Place on bioprobe test plate
• Add reagents
• Take a measurement

Figure 3.2.6. The biotube arm.

5. It contains valves at each end with separate valve for water sampling
   or system emptying.
6. Collector plates are available in different materials, specific to the users'
   requirements, ranging from stainless steel to galvanized mild steel.

Laboratory Rig Systems

Laboratory rig systems can be used in the laboratory to provide a more
detailed study of biofilm formation or to develop biocide strategies. These
systems are circulating rigs containing small volumes of real or simulated
process water. Such systems are more sophisticated instruments than onsite
units and provide a means of controlling parameters under investigation. This
device also has disadvantages, in that it does not reflect the scale of biofouling
associated with the real system. Also the geometry and hydrodynamics
represented by the device may be unrealistic. Most laboratory systems are
based on the 'feed and bleed system' (Gilbert and Herbert 1987; King 1995).
Thus the major part of the liquid volume of the rig is circulated with a
continuous addition of a small volume of fresh medium (feed) and removal of
an equal volume of spent medium and culture (bleed). These systems can also
be employed on a continous batch feed system (Videla 1996).

Tubular Geometry Biofilm Monitoring Device

Tubular geometry biofilm monitoring devices allow the formation of
biofilms on the inside surface of tubular pipe sections. Generally, three
types of devices have been cited (Cloete *et al.* 1992). These include:

1. Ported tube—This device consists of a small diameter pipe fitted with
   sampling ports containing removable studs (Cloete *et al.* 1992).

2. Sectioned tubes—These are sections of a tube that may be contained within another assembly. Some devices are available that have a scored outer surface allowing sections to be broken off or sectioned, using a pipe cutter, for analysis. This system however, does come with a particular disadvantage, with regards to microscopic analysis of the biofilm, as the curved surface prevents accurate visualisation and interpretation.

3. Monitored tube—This equipment allows the measurement of the heat resistance and fluid frictional resistance. The rate and extent of biofouling on a surface is directly correlated to any changes in these parameters.

Another device cited in the literature involves the use of sections of pipe, joined by compression fittings, with slides incorporated within the sectioned pipes (Percival *et al.* 1998a, b and c). This enables biofilm development and characterization in a 'real system', whilst enabling true parameters that effect biofilm development within water systems to be measured. Other devices using slides have been documented by Pedersen (1982).

CHOICE OF MONITORING DEVICE

The choice of device to detect and monitor biofouling and biofilm formation is dependant upon economic constraints and the complexity of the engineered system being monitored. Therefore, a number of considerations need to be taken into account within a monitoring programme (Flemming and Geesey 1991):

1. The water quality in the system.
2. The design of a facility for testing the impact of proposed treatment programmes on various components and materials.
3. Determination of biofouling/biocorrosion under stagnant, low flow and high flow rates.
4. Monitoring various locations such as valves and welds, known to be affected by corrosion.
5. Provide for regular visual inspection by removing samples for analysis.
6. Provide for the expansion of side-stream devices to include additional fouling equipment.

EXAMINATION OF BIOFILMS ON SAMPLING DEVICES

Microscopy

For the study of biofilm formation and biofouling, the quickest method for analysis is by microscopy. This method can indicate the presence of micro-organisms upon a surface, ranging from bacteria to algae. A major drawback is that it does not provide a means of identifying adhered microbes, however, it will provide an indication of the degree of contamination.

Direct enumeration of sessile cells, using epifluorescence microscopy, has often been unable to distinguish between living and dead cells, although several methods have recently been developed, to combine direct microscopy with a viability test to quantify total living cells. A number of microscopes exist for the analysis of biofilms including epifluorescene to the atomic force microscope (AFM). Epifluorescence involves the use of acridine orange (Daley and Hobbie 1975; King 1995) or DAPI (4′6-diamino-2-phenylindole) to stain bacteria. These stains which are DNA specific have been used to differentiate between both living and dead cells (Atlas and Bartha 1987; Wolfaardt et al. 1991). Other methods include the use of iodonitrotetrazolium (INT) formazan, developed by Zimmerman et al. (1978), and the nalidixic acid method of Kogure et al. (1979). The addition of DNA gyrase as an inhibitor to nalidixic acid provides additional information on the status of cells. This antibiotic agent prevents cell division, but not carbon processing in the bacteria. Elongation of treated cells provide osmotically stable cells, although, it is not possible to demonstrate leakage of intracellular components from the treated cells. Goss et al. (1964) has found that elongation of treated cells is a function of nalidixic acid concentration in metabolically active cells. The elongated cells may then be enumerated and measured microscopically to determine viable and metabolically active cells.

The redox dye 5-cyano-2,3-ditolyl tetrazolium chloride (CTC) has recently been introduced for the enumeration of metabolically active bacteria (Rodriguez et al. 1992). CTC is readily reduced to an insoluble, highly fluorescent and intracellular CTC-formazan via an electron transport system. Fluorescence emission of CTC-formazan has been found primarily in the red region, with excitation via long-wavelength UV light (Yu et al. 1995). Severin and associates (1985) have proposed that CTC reduction occurs on the cell surface, but it is most often used for intracellular determination of oxidative enzymes. They suggest that the tetrazolium salt reduction takes place under the influence of membrane-bound redox enzymes known to exist on the cell surface. Combined flourescence staining procedures provide a more sensitive and rapid approach for the enumeration of viable microorganisms. By linking these techniques to image analysis systems, much of the subjectivity inherently linked to plate-count methodology may be removed.

Other methods used to measure total counts of bacteria involve the use of microscopic counting chambers such as the Neubauer hemocytometer. However, the viability of microbiological cells cannot be assessed by this method. In addition, the use of light microscopy for the analysis of biofilms has been reported (Walker and Keevil 1994).

Scanning electron microscopy (SEM) is a technique often used for the analysis of biofilms within industrial processes, to characterize the biofouling and biocorrosion phenomena (McCoy et al. 1981; Percival et al. 1997). The visualization of biofilms using this technique, however, is classed as unrepresentative of the in situ environment due to the sample preparation. The dehydration with alcohol is known to lead to artefacts in the biofilm. Also the correlation of biofilm formation evident between SEM micrographs and a sessile total viable count is cited as being very low (Wolfaardt 1990). With the development of the environmental scanning electron microscope, hydrated biofilm samples may be analysed in their natural state avoiding the development of artefacts due to laborious samples preparation (Wagner et al. 1992).

More 'state of the art' microscopes are being used due to their potential in analysing biofilms and microbial corrosion. These include the atomic force microscope (AFM) (Steele et al. 1994), scanning tunnelling microscope (STM) and the confocal laser microscope (CLSM). These instruments are expensive and are presently confined to the research rather than the industrial workplace.

Microscopic examination may not be practical for monitoring industrial processes, particularly if large amounts of debris are associated with these waters. This may lead to the formation of large amounts of biofouling, and therefore, extensive biofilm coverage.

Indirect Measurement of Biofilms and Biofouling

Indirect measurements used to detect the presence of biofouling/biofilms within the industrial processes include culture techniques, activity measurement and estimation of biomass. Infra-red spectroscopy has been used to indicate the extent of biofilm formation (Santos et al. 1991). Other techniques being applied in industrial water systems to measure biofouling involve the detection of changes in either pressure across a test section or heat transfer from a heated surface (Characklis et al. 1986).

Culture Techniques

Microorganisms are routinely isolated and identified by their ability to grow on various liquid or solid media. Media used in the laboratory for both isolation of bacterial and fungal cultures are numerous and can be located elsewhere (Videla 1996). The enumeration of heterotrophic organisms within water systems is necessary to establish a risk assessment of the likelihood and the extent of biofouling and corrosion.

In order to analyse the microbial content of biofilms, they are removed from the surface (coupons or studs) by scraping into sterile saline or double distilled water. The biofilm is resuspended by vortexing or sonicating to disperse clumps of bacteria. The suspension is then plated out using serial dilution to extinction. Depending on the industrial process, microbial enumeration and identification is carried out using different selective media, incubation times and temperatures. Plate counts are generally taken after 48 or 72 h (Greenberg *et al.* 1985; Gibbs and Hayes 1988), but due to the evidence of very slow growing organisms it is often necessary to incubate plates for longer periods of time.

Sulphate reducing bacteria (SRB) are general indicators of corrosion or its initiation (Videla 1996). It is important to quantify the number of microorganisms present for biofouling and corrosion. When the level of SRBs is 10^4 cm^{-2} it can be considered as a clear indication that corrosion problems will soon appear. SRB cultures take up to 10 days to develop, therefore, the development of more rapid techniques are required. The rapid techniques include ATP analysis, radiorespirometry and immunological techniques and they will be discussed in more detail later.

One of the simplest methods to rapidly detect whether or not microorganisms are present in a specific environment and estimating levels of contamination is the use of a dipstick. A dipstick consist of a sterile plastic strip coated with a layer of nutrient agar. The dipstick is dipped into the fluid to be tested and resealed in its container. The sticks are incubated in a warm place for 24 hours and the number of bacteria and fungi can be determined by using a calibration chart supplied by the manufacturer. The advantage of this method is that non-specialists can use it, quickly and reliably, onsite. In addition, the dipsticks can be used on surfaces as tough plates. If the microorganisms are required for identification, the stick can easily be sent to a microbiology laboratory. However, the limitation of this procedure is that very dilute cell suspensions are not reliable.

Estimation of Biomass

The determination of biomass within biofilms is very difficult to measure. The ideal techniques which are presently available to measure biomass include adenosine triphosphate (ATP) measurements, immunological techniques and gene probes (Videla 1996).

ATP is a useful tool to estimate the extent of biomass within industrial process water. As ATP is present within all bacteria, this will enable accurate determination of biofouling or planktonic biomass. ATP rapidly decreases after cell death, therefore, its concentration will relate to living biomass (Atlas and Bartha 1987). The measurement of ATP is based on the production of bioluminescence using the luciferin/luciferase system (Challinor 1991). This technique is widely used for testing industrial process water to estimate the microbial levels. The method, however, does require both skilful and careful interpretation. The advantage of this assay is the very short time scale for results, allowing for corrective action to be taken before irreversible effects occur. The ATP assay has also been used within the oil field (Prasad 1988).

A rapid ATP bioassay is available to identify biofilms within cooling water systems (Chalut *et al.* 1995). This involves the use of an ATP sampling pen (Bioscan™) consisting of four parts (Figure 3.2.7). First, a sampling stick located at the bottom of the pen, is coated with a cell lysing agent to extract the ATP. This part is dipped into the biofilm sample for one second. The second part is the cuvette, located in the middle of the pen, which acts as a mixing chamber for the sample and the reagents. The cuvette also allows light to be emitted for measurement by a photometer. The third part is a cap which houses the luciferin/luciferase reagent, and finally, a button which pushed through the cap seal to release the luciferin/luciferase enzyme into the cuvette prior to mixing.

Immunological methods have also been employed as indicators of biofilm growth (Pope and Zintel 1988). These techniques include agglutination (Abdollahi and Nedwell 1980), immunodiffusion (Postgate and Campbell 1963) and immunoflourescence (Smith 1982). Enzyme labelled antibiotics can provide a means of identifying different strains of bacteria. Enzyme-linked immunosorbent assays (ELISA) provide a rapid and more sensitive technique for the detection of bacteria, particularly slow growing SRBs (Bobowski and Ndwell 1987; Gaylarde and Cooke 1987).

The use of gene probes are also showing promise for the detection of biofilms, particulary when low numbers of specific bacterial species are present in a mixed microbial population. The use of nucleic acid gene probes are being applied to SRBs within oil field fluids (Westlake *et al.* 1993). Also, the use of chromosomal DNA hybridization techniques and reverse sample genome probing have allowed for the identification of different bacteria in a sample (Videla 1996).

ON-LINE MONITORING

On-line monitoring systems used for the detection of biofouling would be ideal, however, no system is commercially available to study biofilm formation in 'real time'. Rapid progress has been made in the development of sensors to assess both corrosion and biofouling (Winters *et al.* 1993; Stokes *et al.* 1994). Automatic image analysis, Fourier transform infrared spectroscopy, quartz crystal microbalance and electrical impedance spectroscopy are also being developed into on-line biofouling monitors and detection systems.

New techniques are needed for direct assessment of biofilm formation in industrial processes. Challinor (1991) listed the requirements for such

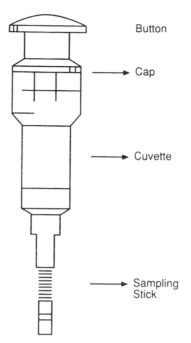

Figure 3.2.7. An ATP sampling pen.

a test system, it must be rapid; able to measure and differentiate between planktonic and sessile microbial populations; simple, requiring minimum facilities and unspecialized staff; accurate; reliable; reproducible; cost effective; and robust, to enable use within an industrial setting.

Flemming (1991) has also provided requirements of an ideal monitoring system and includes:

- Monitoring of biofilm formation on representative surfaces.
- Monitoring of the microbial quality of feedwater and all pretreament chemicals.
- Monitoring of plant performance data to enable correlation to be established for both predicting and evaluating biofouling.

An on-line monitoring system that may be used to study biofilm activity in cooling water is an electrochemical probe called the BioGeorge™. This device is able to monitor changes in electrochemical reactions induced by biofilms on stainless steel electrodes (Licina *et al.* 1992; Licina and Nekoksa 1993; Licina *et al.* 1993).

CONCLUSION

It is clear that no single system is able to satisfy all criteria which are necessary for the complete monitoring of biofouling in industrial processes and waters. Despite this, it is essential that monitoring and detection of biofilms within industrial waters is carried out to reduce the impact of biofilm formation on plant operations. This in turn, will reduce treatment procedures by the use of overdosing or ineffective concentrations of biocides. The most promising monitoring system is that of the Bioprobe and Biotube, which will provide immediate *in situ* measurements, eliminating long time-consuming procedures. However, these will not provide the isolation and identification of biofilm microorganisms.

All detection methods within a complete industrial process must provide the maximum amount of data to enable accurate interpretation to detect biofouling during its primary development. This will save time and ultimately reduce operating costs.

ACKNOWLEDGEMENT

I would like to thank Dr David Webber of Ashland Ltd. for providing information on the Ashland biofilm monitor and the bioprobe system.

REFERENCES

Abdollahi, H and Nedwell, DB (1980) Serological characteristics within the genus *Desulfovibrio*, Antonie van Leeuwenhook. *Journal of Microbiology and Serology* 56:73–78.

Anon. (1991)The control of legionellosis including legionnaires' disease. Health and Safety Series booklet, HS(G) 70, HMSO.

Atlas, RM and Bartha, R (1987) Evolution and structure of microbial communities. In: Microbial Ecology: Fundamentals and Applications, 2nd edn, The Benjamin/ Cummings Publishing Co., Menlo Park, CA, pp. 61–98.

Bobowski, S and Ndwell, DB (1987) A serological method using a microElisa technique for detecting and enumerating sulphate-reducing bacteria. In: Industrial Microbiological Testing (Eds. Hopton, JW and Hill, EC), Blackwell Scientific Publications, Oxford, UK, pp. 171–180.

Bott, TR and Miller, PC (1983) Mechanisms of biofilm formation on aluminium tubes. *Journal of Chemical Techniques and Biotechnology* 33B:177–182.

Bott, TR (1992) Introduction to the problem of biofouling in industrial equipment. In: Biofilms Science and Technology (Eds. Melo, LF, Bott, TR, Fletcher, M and Capdeville, B), Kluwer Academic Publishers, Boston, pp. 3–12.

Challinor, CJ (1991) The monitoring and control of biofouling in industrial cooling water systems. *Biofouling* 4:253–266.

Chalut, J, D'Arise, L, Bodkin, PM and Stodolka, C (1995) Identification of cooling water biofilm using a novel ATP monitoring technique and their control with the use of biodispersants. In: Corrosion/88, paper No. 211, NACE International, Houston, TX.

Characklis, WG, Zelver, N and Roe, FI (1986) Continous on-line monitoring of microbial deposition on surfaces. *Biodeterioration* **VI**:427–435.

Characklis, WG (1985) Influence of microbial films on industrial processes. In: Proceedings of an Argentina–USA Workshop on Biodeterioration (CONICET-NSF) (Ed. Videla, HA), Aquatec Quimica S.A., São Paulo, pp. 181–216.

Cloete, TE, Brozel, VS and von Holy, A (1992) Practical aspects of biofouling control in industrial water systems. *International Biodeterioration* **29**:299–341.

Daley, RJ and Hobbie JE (1975) Direct counts of aquatic bacteria by a modified epiflourescence technique. *Limmnology and Oceanography* **20**:875–883.

Flemming, HC and Geesey, GG (1991) *Biofouling and Biocorrosion in Industrial Water Systems.* Proceedings of the International Workshop on Industrial Biofouling and Biocorrosion, Stuttgart, Springer-Verlag, New York, pp. 1–6.

Gaylarde, CC and Cooke, PE (1987) ELISA techniques for the detection of sulphate-reducing bacteria, In: *Immunological Techniques in Microbiology* (Eds. Grange, JM, Fox, A and Morgan, NL), Society for Applied Bacteriology Technical Series No. 24, 231.

Gibbs, RA and Hayes, CR (1988) The use of R2A medium and the spread plate method for the enumeration of heterotrophic bacteria in drinking water. *Letters in Applied Microbiology* **6**:19–21.

Gilbert, PD and Herbert, BN (1987) Monitoring microbial fouling in flowing systems using coupons, In: Industrial Microbiological Testing (Eds. Hopton, JW and Hill, EC), Blackwell Scientific Publications, Oxford, UK, p. 79.

Goss, WA, Deitz, WH and Cook, TM (1964) Mechanism of action of nalidixic acid on *Escherichia coli. Journal of Bacteriology* **88**:1112–1118.

Green, PN, Bousfield IJ and Stones, A (1987) The laboratory generation of biofilms and their use in biocide evaluation, In: Industrial Microbiological Testing (Eds. Hopton, JW and Hill, EC), Blackwell Scientific Publications, Oxford, UK, pp. 99-108.

Greenberg, AE, Trussel, RR and Clesceri, LS (1985) Standard methods for the examination of water and wastewater, 16th edn. American Public Health Association, Washington, D.C.

King RA (1995) Monitoring techniques for biological induced corrosion. In: Bioextraction and Biodeterioration of Materials (Eds. Gaylarde, CC and Videla, HA), Cambridge University Press, Cambridge, p. 271.

King, RA (1996). Monitoring techniques for biologically induced corrosion. In: Bioextraction and biodeterioration of metals (Eds Gaylarde, C and Videla, HA), Cambridge University Press, Cambridge, pp. 271–306.

Kogure, K, Simudu, U and Taga, N (1979). A tentative direct microscopic method of counting living marine bacteria. *Canadian Journal of Microbiology* **25**:415–417.

Licina, GJ, Nekoksa, G and Howard, RL (1992) An electrochemical method for on-line monitoring of biofilm activity in cooling water. In: Corrosion/92, paper No. 177, NACE International, Houston, TX.

Licina, GJ, and Nekoksa, G (1993) On-line monitoring of microbiologically inflenced corrosion in power plant environments. In: Corrosion/93, paper No. 403, NACE International, Houston, TX.

Licina, GJ, Nekoksa, G and Howard, RL (1993) The BioGeorge probe. An electrochemical method for on-line monitoring of biofilm activity. In: Proceedings

NSF-CONICET Workshop on Biocorrosion and Biofouling Metal/Microbe Interactions (Eds. Videla, HA, Lewandowski, Z and Lutey, RW), Buckman Laboratories, International Inc., Memphis, TN, 122.

McCoy, WF, Bryers, JD, Robbins, J and Costerton, JW (1981) Observations of fouling biofilm formation. *Canadian Journal of Microbiology* 27:910–917.

Miller, JDA (1971) Microbial Aspects of Metallurgy. Medical and Technical Publications, Aylesbury.

NACE (1990) Microbially Influenced Corrosion and Biofouling in Oilfield Equipment, TPC-3; 1990 revision T-ID-26. NACE, Houston, TX.

Payer, JH, Dippold, DG, Boyd, WK, Berry, WE, Brooman, EW, Buhr, AR and Fisher, WH (1978) Economic effects of metallic corrosion in the United States. A report to NBS by Batelle Columbus Laboratories, US Department of Commerse, Washington, D.C.

Pedersen, K (1982) Method for studying microbial biofilms in flowing-water systems. *Applied Microbiology* 43:6–13.

Percival, SL, Beech, IB, Edyvean, RGJ, Knapp, JS and Wales, DS (1997) Biofilm development on 304 and 316 stainless steels in a potable water system. *Journal of the Institute of Water and Environmental Management* 11:289–294.

Percival, SL, Knapp, JS, Wales, DS and Edyvean, R (1998a) The effects of the physical nature of stainless steel types 304 and 316 on bacterial fouling. *British Corrosion Journal* 33:121–129.

Percival, SL, Knapp, JS, Wales, DS and Edyvean, R (1998b) Biofilm development on stainless steel grades 304 and 316 in potable water. *Water Research* 32:243–253.

Percival, SL, Knapp, JS, Wales, DS and Edyvean, R (1998c) Biofilm, stainless steel and mains water. *Water Research* 32:2187–2201.

Pritchard, AM (1981) Fouling: science or art? An investigation of fouling and antifouling measures in the British Isles. In: Fouling of Heat Transfer Equipment (Eds. Somerscales, R and Knudsen, K), Hemisphere Publishing Corp., Washington, D.C., p. 531.

Pope, DH and Zintel, TP (1988) Methods for the investigation of under-deposit microbiologically influenced corrosion. In: Corrosion/88, paper No. 249, NACE International, Houston, TX.

Postgate, JR and Campbell, LL (1963) Identification of Coleman's sulphate reducing bacterium as a mesophilic relative of *Clostridium nigrificans*. *Journal of Bacteriology* 86:274–282.

Prasad, R (1988) Pros and cons of ATP measurement in oil field waters. In: corrosion/88, paper No. 87, NACE International, Houston, TX.

Rodriguez, GG, Phipps, D, Ishiguro, K and Ridgway, HF (1992) Use of fluorescent redox probe for direct visualisation of actively respiring bacteria. *Applied and Environmental Microbiology* 58:1801–1808.

Ruseska I, Robbins, J and Costerton, JW (1982) Biocide testing against corrosion causing oil-field bacteria helps control pigging. *Oil and Gas Journal* 80:253–264.

Sanders, PF and Hamilton, WA (1986) Biological and corrosion activities of sulphate-reducing bacteria in industrial process plant. In: Biologically Induced Corrosion (Ed. Dexter, SC), NACE International, Houston, TX, pp. 47–52.

Santos, R, Callow, ME and Bott, TR (1991) The structure of *Pseudomonas fluorescens* biofilms in contact with flowing systems. *Biofouling* 4:319–328.

Severin, E, Stellmach, J and Nachtigal, HM (1985) Fluorimetric assay of redox activity in cells. *Analytica Chimica Acta* 170:341–346.

Smith, AD (1982) Immunofluorescence of sulphate-reducing bacteria. *Archives of Microbiology* 133:188–121.

Steele, A, Goddard, DT and Beech, IB (1994) An atomic force micrscopy study of the biodeterioration of stainless steel in the presence of bacterial biofilms. *International Biodeterioration and Biodegradation* **34**:35–42.

Stokes, PSN, Winters, MA, Zuniga, PO and Schlottenmier, DJ (1994) Developments in on-line fouling and corrosion surveillance. In: Microbiologically Influenced Corrosion Testing (Eds. Kearns, JR and Little, BJ), ASTM Publications STP 1232, American Society for Testing and Materials, Philadelphia, PA, pp. 1232–1240.

Videla, HA (1994) New trends in biocorrosion/biofouling monitoring techniques. In: Biodeterioration Research 4 (Eds. Llewellyn, GC, Dashek, WV and O'Rear, CE), Plenum Press, New York.

Videla, HA, Bianchi, F, Freitas, MMS, Canales, CG and Wilkes, JF (1994b) Monitoring biocorrosion and biofilms in industrial waters: a practical approach. In: Microbiologically Influenced Corrosion Testing (Eds. Kerns, JR and Little, BJ), ASTM Publications STP 1232, American Society for Testing and Materials, Philadelphia, PA, p. 128.

Videla, HE (1996) Manual of Biocorrosion, CRC Press, Boca Raton.

Wagner, PA, Little, BJ, Ray, RL and Jones-Meecham, J (1992) Investigation of microbiologically influenced corrosion using environmental scanning electron microscopy. In: Corrosion/92, paper No. 185, NACE International, Houston, TX.

Walker, JT and Keevil, CW (1994) Study of microbial biofilms using light microscope techniques. *International Biodeterioration and Biodegradation* **34**:223–236.

Westlake, DWS, Voordouw, G and Jack, TR (1993) Use of nucleic acid probes in assessing the community structure of sulphate-reducing bacteria in Western Canadian oil field fluids, Proc. 12th International Corrosion Congress, NACE International, Houston, TX, 3794.

Winters, MA, Stokes, PSN, Zuniga, PO and Schlottenmier, DJ (1993) Development of on-line corrosion and fouling monitoring in cooling water systems. In: corrosion/93, paper No. 392, NACE International, Houston, TX.

Wolfaardt, GM, Archibald, REM and Cloete, TE (1991) The use of DAPI in the quantification of sessile bacteria on submerged surfaces. *Biofouling* **4**:265–274.

Wolfaardt, GM (1990) Techniques for studying biofouling and aspects influencing biofouling in industrial water systems. MSc Thesis, University of Pretoria, South Africa.

Yu, W, Dodds, WK, Banks, MK, Skalsky, J and Strauss, EA (1995) Optimal staining and sample storage time for direct microscopic enumeration of total and active bacteria with two fluorescent dyes. *Applied and Environmental Microbiology* **61**:3367–3372.

Zimmerman, R, Iturriaga, R and Becker-Birck, J (1978) Simultaneous determination of the total number of aquatic bacteria and the number thereof involved in respiration. *Applied Environmental Microbiology* **36**:926–935.

3.3 Control of Biofilms in Industrial Waters and Processes

HANS-CURT FLEMMING[1,2] and THOMAS GRIEBE[1]
[1]University of Duisburg, Dept. Aquatic Microbiology, Geibelstr. 42, 47057 Duisburg and [2]IWW Centre for Water, Dept. Microbiology, Moritzstr. 26, 45476 Mülheim, Germany

INTRODUCTION

Biofouling occurs in very different fields of industrial waters and processes (Flemming and Schaule 1996a), ranging from the production of ultrapure water for electrical and pharmaceutical products (Schaule and Flemming 1997), drinking water (Flemming et al. 1998), membrane water treatment (Ridgway and Flemming 1996) to paper production (Klahre et al. 1997). Of course, the susceptibility to biofouling is very different. Nevertheless, some interesting similarities can be perceived. In most cases, control strategies seem to be based on a medical paradigm: the system has a microbial 'disease' which is to be treated with antibiotics, i.e. biocides. Killing the microorganisms will not solve the problem. In contrary to infected living organisms, technical systems, do not possess any means for disposal of dead microbes. It may be more important to remove the biomass from the system rather than killing it and leaving it in place. Consequently, the forces that keep biofilms together are of increasing importance because they have to be weakened in order to disperse the biomass so that it can be flushed away.

Another important aspect in biofouling of industrial processes is that 'biofouling' is operationally defined as biofilm development that exceeds a given 'threshold of interference'. Keeping biofilm development below that threshold offers a strategy, which considers handling biofouling as a biofilm reactor in the wrong place (Flemming et al. 1996). Nutrient limitation is an option to limit biofilm development in sensitive areas. If large colonization areas are offered ahead of a protected system, biofilms will develop, sequestering dissolved nutrients. This immobilized biomass will be in a place where it can be easily handled, in contrary for example, inside a heat exchanger or a membrane module. Combined to effective monitoring techniques, a biocide-free anti-fouling-strategy can be designed.

Industrial Biofouling. Edited by J. Walker, S. Surman, J. Jass
©2000 John Wiley & Sons Ltd

Regardless of the system, control of biofouling generally can be divided into three steps:

(i)   detection, i.e. the verification of the diagnosis,
(ii) sanitation of the actual fouling situation and
(iii) prevention of further fouling events.

DETECTION OF BIOFILMS

As mentioned before, the presence of biofilms in a technical system is usually detected indirectly. Operational parameters usually represent the symptoms. In ultrapure water systems, biofouling is recognized by microbial contamination or function failure of the products. In heat exchangers, heat transfer decreases while drag resistance and pressure drop increases. In membrane technology, both hydraulic membrane resistance and feed-brine pressure drop increase. Apart from direct microbial contamination, the effects of biofouling are indirect and non-specific. Thus, recognition in practice, frequently follows the same pattern: a problem in plant performance or product quality occurs; all common countermeasures are applied; if they do not work, and there is no better explanation, the problem is attributed to biology. In most cases this is correct, but it must be verified. Therefore, the correct sampling strategy is crucial (Schaule *et al.* 2000).

Sampling

As soon as biofouling is suspected, water samples are taken and investigated microbiologically. Different methods are applied which often give different results. Cultivation methods detect only those organisms which are able to multiply on the growth media used. The majority of bacteria occurring in natural and technical waters do not grow on common microbiological media (Colwell 1991). According to cultivation temperature and time, the numbers of colony forming units (cfu) can vary by some orders of magnitude.

Direct microscopy counting methods using DNA specific dyes, such as acridine orange, give the number of cells present in a sample but cannot discriminate between viable and non-viable cells. Cell viability is possible to detect by the use of a redox dye, which is reduced to a water-insoluble fluorescent formazan product via the microbial electron transport system (Schaule *et al.* 1993a).

A more fundamental problem in attempting to detect biofouling by sampling of the water phase is that there is no correlation between numbers of suspended cells (detected by water sampling) and the site or extent of biofilm formation. Although biofilms contaminate the water phase, they do so discontinuously. Cells can be eroded, leaving the biofilm active (Szewzyk and Schink 1988), and parts of the biofilm can slough off, leading to episodes of high cell numbers in the bulk liquid phase. Even with low numbers in the water phase, considerable biofilm growth can occur, as demonstrated with microbially contaminated ion exchangers. Although, 5–8 cfu ml^{-1} were found in the process water, sampling of the exchanger bed revealed cell numbers as high as 10^5 cfu ml^{-1} in the wet exchange resin (Flemming 1987). Similar observations are reported by other authors, which showed that the number of sulphate reducing bacteria in the water phase was completely unrelated to the number of biofilm cells (Costerton and Boivin 1987). The concentration of cells released from biofilms is usually diluted in the body of water. However, a frequently observed phenomenon in situations of stagnation, is that cells can concentrate. This has led to the opinion that biofouling occurs preferentially in stagnant water. This effect is not due to the fact that the cells grow better under stagnant conditions. On the contrary, cells grow better under flowing conditions; this is why fermenters are vigorously stirred. In stagnant water, the bacteria accumulate above the biofilm, while they do not in flowing water. Cell numbers in the water phase can therefore be used to identify grossly contaminated sections of a water system. This is very helpful in order to divide large water systems into smaller compartments, e.g. filters, reservoirs, pipe sections and other components. The exact location and extent of biofilm growth, however, is not reflected by water phase analysis. As a consequence, proper sampling of biofilms must be carried out on the surfaces in question or on representative surfaces using appropriate methods (Mittelman and Geesey 1987).

In practice, it will be necessary to get access to fouled surfaces in order to sample the biomass for further analysis. This may be difficult because most water systems are not designed for access to inner surfaces, a prerequisite for proper sampling of biofilms. Good candidates for fouling are O-rings, seals, crevices and valves. State of the art is still scratching the biofouling material from defined surface areas to analyse them. If this is not possible, test surfaces may be placed into the system and removed after a given period of time. In severe biofouling cases, it may be necessary to remove parts of the system in order to get access to the sample surfaces. An example of this is the 'module autopsy' as carried out in membrane biofouling cases (Schaule et al. 1993b).

Analysis of Biofilms

Field Methods

Biofilms in industrial systems have usually reached the plateau phase and, are in most cases, visible with the bare eye. Thus, optical inspection provides a first hint of fouling. In technical environments, biofilms tend to display a slimy consistency, which can be detected by wiping, even a thin layer can be made visible by using a white tissue for wiping. In case of doubt about the biological origin of a deposit or layer, it is useful to take a small amount and put it over a lighter until it smoulders. A smell of burnt protein is characteristic for biological material. Presently, a test kit with a simple dye reaction is under development to deliver reliable information about the presence or absence of biofilms and to semi-quantify them. Conclusive detection of biofilms is carried out in the laboratory.

Laboratory Methods

The chemical analysis must take into account the composition of biofilms. The main component of biofilms is water (up to 90%). The microorganisms can make up less of the organic content than their extracellular polymeric substances (EPS), also called 'slime substances'. These consist mainly of polysaccharides, proteins, nucleic acids and lipids, and while their proportion can vary considerably, the polysaccharides can represent only a minor fraction of the EPS. Due to the adhesive properties of the EPS matrix, particles can be trapped, accumulate and represent the main proportion of a deposit. The deposits at any given site are cells 'glued together' and entrapped by the EPS of the biofilm. Thus, overall parameters such as the water content of a deposit can be indicative of biofilms, although it is non-specific. The same is true for the organic carbon content of deposits. Table 3.3.1 summarizes some parameters suitable for the identification of biofilms.

The presence of adenosine triphosphate (ATP) or respiration activity clearly indicate living organisms. The other parameters are also characteristic of biofilm formation and are mostly to quantify the biomass. It is advised to use more than one parameter for the characterization of a biofilm. If the occurrence of a problem can be related to the actual presence of biofilms and a correlation between those two is established, the diagnosis 'biofouling' is verified.

SANITATION

As indicated in the introduction, 'disinfection' is the most common countermeasure in biofouling. This can lead to very disappointing results,

Table 3.3.1. Examples for biofilm parameters (after Flemming and Schaule 1996b)

Parameter	Method/Reference
Water content	24 h, 110° C
Organic carbon	TOC, COD, incineration loss
Protein	Bradford (1976)
Carbohydrates	Dubois (1956)
DNA	Palmgren and Nielsen (1996)
Lipids	Geesey and White (1990)
Muramic acid	Geesey and White (1990)
Polyhydroxybutyrate	Geesey and White (1990)
Total cell number	Meyer-Reil (1978)
Colony forming units	various standard methods
ATP	Young-Bandala and Kajdasz (1983)
Hydrolase activity	Obst and Holzapfel-Pschorn (1990)
Respiration activity	Schaule et al. (1993)
Indolacetic acetic acid production	Bric et al. (1991)
Catalase activity	Line (1983)

TOC, total organic carbon; COD, chemical oxygen demand.

as killing the organisms is often not the solution to the problem. Technical systems generally cannot be run sterile (apart from pharmaceutical or electrical water supply systems that are kept aseptic with great effort), re-infection of the 'disinfected' system is assured. New cells will encounter dead biomass that is readily biodegradable and will support rapid microbial regrowth. The efficacy of biocides against biofilm organisms is known to be less when compared to the same organisms suspended in the water phase (LeChevallier et al. 1988). Keevil et al. (1990) reported that coliform organisms in a biofilm survived prolonged exposure to 12 ppm free chlorine. Living biofilms have been found on the inner walls of disinfected pipes (Exner et al. 1983). Unfortunately, most data on biocides relate to suspensions of test organisms. Although this evaluation of biocide efficacy has the advantage to be reproducible, it does not reflect the activity against biofilms. It is very difficult to obtain biofilms that can be used as reference systems with sufficient similarity to 'natural' biofilms.

The use of biocides can be justified in the framework of a combined strategy to minimize the biofouling potential. However, the considerations discussed must be kept in mind. The varieties of biocides available have grown (overviews: Payne 1988; Paulus 1993). The selection of a biocide suitable for a given system requires experience and should be supported by laboratory screening.

The use of chlorine will be increasingly limited by tighter environmental regulations. Thus, ozone offers some interesting advantages. Ozone produces less toxic by-products and it has been shown to be effective against biofilms (Bott 1990). Ozone weakens the biofilm matrix and thus, facilitates the removal of biomass by shear forces (Kaur *et al.* 1992). A drawback is that ozone generation costs about 4 times more than that of chlorine. A potential problem in the application of oxidizing biocides is that refractory organic substances such as humic acids, can become biodegradable.

Removal of Biofilms from Surfaces

If an oxidizing biocide is used, it will weaken the biofilm matrix as a side effect. This is the case for chlorine which is consumed by reacting with extracellular polysaccharides and proteins. Then, the biofilm can be removed easily from the surface. In this case, it is irrelevant whether the cells are dead or not, since the 'side effect' of decreasing matrix stability is much more important. Some biocides, however, can have the opposite effect. Formaldehyde is used in microbiology as a fixation reagent, cross-linking proteins. Reports from practice have found that formaldehyde disinfection can make the problem worse (Exner *et al.* 1987).

If a system is to be cleaned, the binding forces between the biofilm and the support, and the cohesive forces between the EPS matrix molecules, have to be overcome. Although biofilms can be wiped off easily, they can withstand the shear forces of running water. This is the reason for finding biofilms in water pipelines with very high water velocity (Characklis 1990). Although they become thinner, they are more stable. Another choice for cleaning surfaces is ultrasonics and it has been successfully used in cleaning medical devices, dentistry instruments, heat exchangers and others. Although it has been applied in undefined intensity, ultrasonic energy seems to be a feasible method for cleaning of surfaces from unwanted biofilms. However, it has to be considered that ultrasonics may damage the surface on which the biofilm is growing (Zips *et al.* 1990).

Cleaning strategy.

Killing the cells as effectively as possible and leaving them in place will not solve the problem. In general, a cleaning strategy will include the following two steps:

(i)   Weakening of the biofilm matrix by chemicals such as oxidants like chlorine, ozone, hydrogen peroxide, peracetic acid or others; by alkaline treatment, tensides, enzymes (Wiatr 1990), complex forming substances (Turakhia *et al.* 1983) or by biodispersants (Schmidt 1983; Jaquess 1994). The latter are based on polyethyleneglycol and are believed to weaken the interactions in the EPS matrix as well as the interaction between biofilm and support material. A combination of various agents may increase the efficacy. However, it is important to establish some testing system for quantifying the level of biofilm removal.

(ii) Removal of the biofilm by mechanical forces such as rinsing with water, air, steam, or a combination, the application of sponge balls, brushing, or ultrasonics.

The role of cohesive forces in biofilms

Step (i) in particular seems to be approached completely empirically in practice and there is no real scientific support. The key factor for the dispersion of biofilms is weakening the cohesion forces that keep biofilms together. It is these forces that must be overcome by cleaning, be it mechanically or chemically.

The mechanical stability of microbial aggregates represent a very important parameter. In biofilm technology, e.g. fixed bed or airlift reactors, the performance of the processes can be crucially dependent of the mechanical stability of the biofilm on the support material. If the shear forces exceed the cohesive forces, 'sloughing off' occurs, leading to serious interference of the process which depends on biofilm stability. The reasons for occasional instabilities in biofilm structure are known only by empirical observations. For example, an imbalance of nutrient composition may lead to excessive EPS formation and sloughing off. Biological waste-water treatment is performed largely by the activated sludge process, where the microorganisms are organized in flocs that are formed by EPS. Settling of the flocs is crucial for successful waste water treatment because in this step biomass is removed from the water stream. It depends strongly on floc size and density, which, in turn, depends on the cohesive forces in the aggregate, performed by the EPS.

In the case of unwanted biofilm phenomena such as biofouling or biocorrosion, the removal of the biofilm requires the forces that keep the matrix together and on the surface, to be overcome. Chemicals used for this purpose include oxidizing agents such as chlorine or hydrogen peroxide or non-oxidizing agents such as biodispersants. Thus, for both wanted and unwanted microbial aggregates, mechanical stability as caused by the cohesive forces of EPS molecules and, in the case of biofilms, also by the adhesive forces of EPS to the support.

Both cohesive and adhesive forces are a result of weak interactions and not of covalent bonds. In principal a very simplified view is that three basic cases of weak physico-chemical interactions have to be considered as listed in Table 3.3.2.

Hydrogen bonds are preferably formed by hydroxyl groups. Polysaccharides contain large amounts of these groups. Proteins also contain

Table 3.3.2. Cohesion forces in biofilms: weak physico-chemical interactions of EPS molecules

Interactions	Effects	Examples
Hydrogen bonding	Hydrogel formation	–OH groups in polysaccharides, proteins
Electrostatic interactions	Cation bridging, ion exchange, metal ion sorption	–COO–, phosphate, sulphate, amino groups, induced dipoles
van der Waals interactions	Micelle and membrane formation	Lipids, lipoproteins, lipopolysaccharides

groups capable of hydrogen bonding that are responsible for their tertiary structure and water binding as in hydrogels. The binding energy ranges between 10–30 kJ Mol^{-1} and the distance between the groups is 0.27–0.31 nm. This kind of binding can be disturbed by 'chaotropic agents', such as urea, tetra methyl urea, sodium rhodanide or guanidine hydrochloride, which change the structure of hydrates.

Electrostatic interactions are performed by charged functional groups in macromolecules. Among them are carboxyl, phosphoryl, sulphate, sulfhydryl, sulphide groups as anionic sites and amines as cationic sites. Extracellular polysaccharides such as alginates are bridged by divalent cations (e.g. Ca^{2+} or Mg^{2+}). Proteins contain all of the listed groups, therefore direct electrostatic interactions, as well as, bridging by divalent ions plays an important role in their cohesion. The binding energy ranges between 12–29 kJ Mol^{-1} and the distance between the groups is approximately 0.3 nm. Electrostatic interactions are disrupted by pH changes such as acid and caustic rinsing, changes in ionic strength of the medium such as salt shock, and by the action of complexing substances.

Van der Waals interactions are mainly active among hydrophobic regions of proteins and between hydrophobic surfaces and macromolecules. They are the weakest form of the weak interactions with a binding energy of about 2.5 kJ Mol^{-1}. The distance of this kind of bond is 0.3–0.4 nm. Van der Waals forces can be broken by surface active substances such as tensides.

As the name indicates, all of these interactions are very weak compared to covalent bonds. The binding force, however, increases by the number of interacting groups per molecule. Assuming that an EPS macromolecule carries 10^6 possibly interacting sites with only 10% of them actually involved, the binding force per molecule is magnified by a factor of 10^5. In addition, the resulting tertiary structure, such as random entanglement or the formation of helices, also contributes to cohesion. Different interactions can be active in an EPS matrix at the same time. Therefore, a considerable number of groups can be involved to different degrees, providing cohesion even in the presence of substances that may interfere strongly with one of the forces, but not all. From a standpoint of microbial ecology, this combined effect, together with its poor biodegradability provides a high level of protection for the biofilm.

Enzymes

The experience of the authors in the application of enzymes was not encouraging. This may be due to the specificity of enzymes, which is directed against certain bonds that may not be present in all matrix forming molecules. Thus, microbial strains which produce EPS that the enzymes are not effective against gain a selective advantage. Mechanical cleaning can select for those species which form the most resistant EPS matrix, as has been reported in biofouling problems with the techniques to recover ocean thermal energy (Nickels *et al.* 1981).

Efficacy Control

In practice, the effect of a cleaning measure is assessed by the recovery of process parameters. However, this is an indirect method and not well suited to assess success. For example, the biofouling layer on a reverse osmosis membrane was removed to about 80% by chemical treatment. This led to a substantial improvement of the process parameters. However, the remaining 20% offered optimal conditions for the regrowth of biofilm bacteria. Thus, after a very short time, the old situation was re-established, which led to the well known 'saw tooth curve' (Flemming *et al.* 1994).

Since biofouling is a biofilm problem and it is related to surfaces, the surface has to be sampled properly for effective control. Usually, this is achieved either by removing deposit material from a surface or by exposition of test coupons which are analysed after a given time e.g. 'Robbins device' (McCoy *et al.* 1981). More suitable approaches include integration of test devices which indicate biofilm growth at inaccessible sites on line, non-destructively and in real time, and mobile systems which allow representative and accessible samples surfaces. Some options are listed in Table 3.3.3.

PREVENTION

The most common practice to prevent biofouling is still the addition of biocides. This is usually performed by continuous or intermittent dosing of

Table 3.3.3.   Some options to biofilm monitoring

Principle	References
Integrated test devices	
• Heat exchange resistance	Roe *et al.* (1994)
• Fluid frictional resistance	Roe *et al.* (1994)
• Quartz crystal microbalance	Nivens *et al.* (1991)
• Surface acoustic waves	Ballantine and Wohltjen (1989)
• FTIR–ATR	Schmitt and Flemming (1998)
• FTIR transmittance	Schmitt and Flemming (1996)
• Light reflectance	Tamachkiarowa and Flemming (1996)
• Microcalorimetry	Sand (1987)
• Differential turbidity measurement	Klahre and Flemming (1998)
Mobile systems for surface scanning	
• FTIR–ATR	Schmitt and Flemming (1998)
• Light reflectance	Tamachkiarowa and Flemming (1998)
• Ultrasonics	Zips *et al.* (1990)

FTIR–ATR, Fourier transform infrared–attenuated total reflection spectroscopy.

biocides. However, this strategy is not always successful, because it selects for biocide-tolerant organisms, the biocides are corrosive or incompatible to other conditioning agents, or the biocides represent a contaminant to the waste water which has to be treated separately causing increasing costs.

Methods which rely on the prevention of adhesion have been studied extensively. However, there is no surface which will not be microbially colonized eventually, as extensive studies in naval biofouling research has shown. It seems that it would be more successful to find ways for easy biofilm removal. This may be achieved by using suitable conditioning films (Busscher *et al.* 1995) or antifouling surfaces (Cooksey and Wigglesworth-Cooksey 1992), which facilitate the breakdown of the bonds between biofilm and substratum. Slowest microbial adhesion and lowest adhesion forces are reported where surface tension ranges between 23–27 mJ m^{-2}; below and above these values, the adhesion rate is higher (Dexter 1976; Baier 1980). Copper is widely believed to be bactericidal and copper plating belongs to the first anti-fouling studies in naval history. However, colonization of copper surfaces is a well known phenomenon (Exner *et al.* 1983) and once covered with a primary layer, even copper-sensitive organisms can survive in a biofilm on a copper surface. Another so-called biocidal element is silver, to which the 'oligodynamic effect' is attributed. In fact, most microorganisms are killed by very low concentrations of silver ions ($< 10 \ \mu g \ L^{-1}$) and silver is not toxic to humans. Unfortunately, microorganisms develop a tolerance to silver within a few weeks allowing them to multiply in the presence of more than 1 mg L^{-1} of Ag$^+$ (Flemming 1991).

Good housekeeping

Effective prevention of biofouling can be achieved by a combined strategy. It should be based on a 'clean system philosophy', including all surfaces in contact with water. It includes the minimization of biodegradable organic substances, easy access to surfaces that must be kept clean, surfaces which are easy to clean and have none or very few crevices, fittings, edges, niches and other sites predestined for biofilm growth. Ultra-pure water technology provides many experiences and solutions for less pure water systems. The elements of good housekeeping can be summarized as follows: (i) minimize nutrient concentration; (ii) monitor biofilm development and efficiency of cleaning measures; (iii) strong shear forces; and (iv) frequent cleaning (if possible, mechanical), and easy-cleaning-design.

Living with Biofilms—A Biocide-free Approach to Anti-fouling Strategy

All approaches to prevent biofouling as discussed above are based on killing and cleaning. A completely different strategy can be developed from some basic considerations as mentioned in the introduction. Almost all surfaces in contact with water will have a biofilm. But not all water systems suffer from biofouling. It has been shown that biofilms are involved in the separation process on reverse osmosis membranes right from the beginning. Only after a few hours of operation, a biofilm develops and participates in separation (Flemming et al. 1994). Biofouling occurs only when biofilm development exceeds a certain threshold of interference (Figure 3.3.1).

In technical systems, biofilms usually reach the plateau phase relatively rapidly. What happens is a natural phenomenon: microorganisms settle on surfaces and convert diluted nutrients into metabolic products and locally immobilized biomass. This is technically exploited and is the principle of the biofilm reactor. Biomass accumulates in the reactor and the effluent water will support microbial growth to a lesser degree, including the development of biofilms. Membranes, filter beds, pipe and other technical environments offer large surface areas for colonization, encouraging the same phenomenon, until the biofilm growth exceeds the level of inter-ference. Thus, biofouling can be considered as a biofilm reactor in the wrong place. The main factors which govern biofilm growth, and biofouling potential, are ubiquitous biofilms, the nutrients available and the shear forces. As biofilm formation cannot be avoided without extreme effort (e.g. as performed by the pharmaceutical and microelectronic industries), only nutrients and shear forces remain variable. Shear forces can be increased quite easily in some systems, but not in all. Thus, nutrient concentration

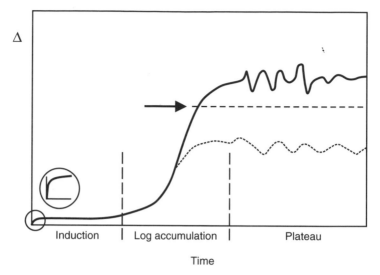

Figure 3.3.1. Schematic development of biofilm accumulation with an arbitrary *threshold of interference* (after Flemming *et al.* 1994).

represents a candidate for limiting biofilm growth. The lower the nutrient concentration, the less biofilm will form. This effect is well known in biofilters, where most biomass accumulates in the areas with the highest nutrient concentration. The biofilm reactor can be 'put in the right place' if positioned prior to a system that is to be protected from biofouling and thus the biofilm will primarily develop in the reactor. In the surrounding areas, biofilm growth will still occur, but below the threshold of interference, if the system is run properly. For example in a cooling water system, (Griebe and Flemming 1996; Flemming *et al.* 1997), river water was treated with a sand bed as a biofilter. Before and after the filter, a reverse osmosis test cell was installed, one operating with raw water and the other with water after the biofilter (parameters given in Table 3.3.4).

Uronic acids are indicative for polysaccharides in the EPS. The protein:uronic acid ratio before the sand filter is 7.7:1 while after the sand filter it is 3.6:2.3. This indicates that the biofilm contains much more EPS under conditions of nutrient depletion. The biodegradable dissolved organic carbon (BDOC) is the biodegradable fraction of the DOC in water. Interestingly, the thickness of the biofilm was only reduced to a third of the original size, but it's structure was different. It appeared that the biofilm formed under nutrient depletion was much less dense than the one formed with raw water. It was sufficient to reach a BDOC of 0.125 mg l^{-1} in order to stay below the threshold of interference. Such a value can be reached easily with common biofilters. The data on microbial

Table 3.3.4. Parameters before and after sandfilter

Parameter	Before sandfilter	After sandfilter
Biofilm		
Thickness	(27.3 ± 3.1) μm	(3.0 ± 0.5) μm
Protein content	77.7 μg cm^{-2}	3.6 μg cm^{-2}
Carbohydrate content	22.5 μg cm^{-2}	2.6 μg cm^{-2}
Uronic acid content	10.6 μg cm^{-2}	2.3 μg cm^{-2}
Water		
BDOC of water	0.325 mg l^{-1}	0.125 mg l^{-1}
Microbial content	1×10^7 cfu ml^{-1}	1.2×10^6 cfu ml^{-1}
Permeate production	65% after 10 days	>95% over entire test period

concentration in the water phase shows that the biofilter does not cause additional microbial contamination but rather removes cells from the process stream.

CONCLUSION

Five expensive mistakes in conventional anti-fouling strategies are identified and can be overcome by the following new approaches:

1. Detection of biofouling occurs by process performance losses or decreased product quality.

   Instead, monitoring of biofouling with early warning capacity is necessary. Such systems are under development.

2. If biofouling is suspected, the water phase is often sampled rather than the surfaces.

   Instead, surfaces have to be sampled and analysed microbiologically and biochemically.

3. Biocides are applied which kill the biomass but do not remove it, giving rise to rapid re-growth.

   Instead, dispersants have to be applied and optimized on a scientific basis, remembering the forces that keep biological aggregates together.

4. The content of biodegradable substances in a system is not considered as potential biomass and not decreased or even increased by biocides and cleaners.

   Instead, biofouling has to be considered as "a biofilm reactor in the wrong place". Nutrient limitation in the right place can limit biofilm growth in areas where it is not wanted.

5. Efficacy control of countermeasures is performed again by process performance losses or decrease of product quality.

Instead, monitoring systems that sense biofilm growth on line, in situ, *in real time and non-destructively are suitable to assess the success of countermeasures and optimize the use of dispersants or other means.*

The authors are well aware that not all biofouling problems can be addressed with the same strategy. However, if the above summarized approaches are applied where suitable, a considerable proportion of biofouling problems can be solved or avoided.

REFERENCES

Baier, RE (1980) Substrata influences on adhesion of microorganisms and their resultant new surface properties. In: Adsorption of microorganisms to surfaces (Eds. Bitton, G and Marshall, KC), John Wiley, New York, pp. 59–104.

Ballantine, DS and Wohltjen, H (1989) Surface acoustic wave devices for chemical analysis. *Analytical Chemistry* **61**:188–193.

Bott, TR (1990) Bio-fouling. In: Fouling of heat exchanger surfaces (Ed. Bohnet, M), Conf. Proc., VDI Ges. P.O. Box 1139, 4000 Düsseldorf 1, 5.1–5.20.

Bradford, MM (1976) A rapid and sensitive method for the quantitation of microgram quantities of protein utilizing the principle of protein dye binding. *Analytical Biochemistry* **72**:248–254.

Bric, JM, Rostock, R and Silverstone, SE (1991) Rapid *in-situ* assay for indolacetic acid production by bacteria immobilized on a nitrocellulose membrane. *Applied and Environmental Microbiology* **57**:535–538.

Busscher, HJ, Bos, R and van der Mei, HC (1995) Hypothesis: Initial microbial adhesion is determinant for the strength of biofilm adhesion. *FEMS Microbiology Letters* **128**:229–234

Characklis, WG (1990) Microbial fouling and microbial fouling control. In: Biofilms (Eds. Characklis, WG and Marshall, KC), John Wiley, New York, pp. 523–633.

Colwell, RR (1991) Viable but non-culturable bacteria in the aquatic environment. *Culture* **12**:2–4.

Cooksey, KE and Wigglesworth-Cooksey, B (1992) The design of antifouling surfaces: background and some applications. In: Biofilms—science and technology (Eds. Melo, LF, Fletcher, MM, Bott, TR and Capdeville, L), Kluwer Academic Publ., Dordrecht, pp. 529–549.

Costerton, JW and Boivin, J (1987) Microbially influenced corrosion. In: Biological fouling of industrial water systems. A problem solving approach (Eds. Mittelman, MW and Geesey, GG), Water Micro Associates, San Diego, pp. 56–76.

Dexter, S (1976) Influence of substratum critical surface tension on bacterial adhesion—*in situ* studies. *Journal of Colloids Interface* **70**:346–354.

Dubois, M (1956) Colorimetric method for determination of sugars and related substances. *Analytical Chemistry* **28**:350–356

Exner, M, Tuschewitzki, GJ and Scharnagel, J. (1987) Influence of biofilms by chemical disinfectants and mechanical cleaning. *Zentralblatt für Bakteriologie u. Hygiene B* **183**:549–563.

Exner, M, Tuschewitzki, G-J and Thofern, E (1983) Untersuchungen zur Wandbesiedlung der Kupferrohrleitung einer zentralen Desinfektionsmitteldosieranlage. *Zentralblatt für Bacteriologie U. Hygiene I. Abt. Orig. B* 177:170–181.

Flemming, H-C, Schaule, G, Griebe, T, Schmitt, J and Tamachkiarowa, A (1997) Biofouling—the Achilles heel of membrane processes. *Desalination* 113:215–225.

Flemming, H-C, Tamachkiarowa, A, Klahre, J and Schmitt, J (1998) Monitoring systems for the detection of biofouling in technical systems. *Water Science Technology* 38:291–298.

Flemming, H-C, Schaule, G, Schmitt, J and Griebe, T (2000) Steps of biofilm sampling and characterization in biofouling cases. In: Investigation of biofilms (Eds. Flemming, H-C, Griebe, T and Szewzyk, U), Technomic Publishers, Lancaster, PA, 1–21.

Flemming, H-C (1987) Microbial growth on ion exchangers—a review. *Water Research* 21:745–756.

Flemming, H-C (1991) Biofouling in water treatment. In: Biofouling and biocorrosion in industrial water systems (Eds. Flemming, H-C and Geesey, GG), Springer, Heidelberg, pp. 47–80.

Flemming, H-C, Schaule, G, McDonogh, R and Ridgway, HF (1994) Mechanism and extent of membrane biofouling. In: Biofouling and biocorrosion in industrial water systems (Eds. Geesey, GG, Lewandowski, Z and Flemming, H-C), Lewis Publishers, Chelsea, Michigan, pp. 63–89.

Flemming, H-C and Schaule, G (1996 a) Biofouling. In: Microbially influenced corrosion of materials—scientific and technological aspects (Eds. Heitz, E, Sand, W and Flemming, H-C), Springer, Heidelberg, pp. 39–54.

Flemming, H-C and Schaule, G (1996 b) Measures against biofouling. In: Microbially influenced corrosion of materials—scientific and technological aspects (Eds. Heitz, E, Sand, W and Flemming, H-C), Springer, Heidelberg, pp. 121–139.

Flemming, H-C, Griebe, T and Schaule, G (1996) Anti-fouling strategies in technical systems—a short review. *Water Science Technology* 34:517–524.

Geesey, GG and White DC (1990) Determination of bacterial growth and activity at solid–liquid interfaces. *Annual Reviews of Microbiology* 44:579–602.

Griebe, T and Flemming, H-C (1996) Vermeidung von Bioziden in Wasseraufbereitungs-Systemen durch Nährstoffentnahme. *Vom Wasser* 86:217–230.

Jaquess, PA (1994) Two approaches to biofilm dispersion. *TAPPI Biological Science Symposium* 233–237.

Kaur, K, Bott, TR and Leadbeater, BSC (1992) Effect of ozone on *Pseudomonas fluorescens*. In: Biofilms—science and technology (Eds. Melo, LF, Fletcher, MM, Bott, TR and Capdeville, B), Kluwer Acad., Dordrecht, pp. 589–594.

Keevil, CW, Mackerness, CW and Colbourne, JS (1990) Biocide treatment of biofilms. *International Biodeterioration* 26:169–179.

Klahre, J, Lustenberger, M and Flemming, H-C (1997) The achilles heel of paper production: microbial problems. Conf. Prod.: Wet end chemistry conference & COST workshop, Gatwick, May 28–29 1997. Pira International, Randalls Road, Leatherhead, Surrey, KT22 7RU, UK.

Klahre, J, Lustenberger, M and Flemming, H-C (1996) Mikrobielle Probleme bei der Papierfabrikation. Teil III: Monitoring. *Das Papier* 52:590–596.

LeChevallier, MW, Cawthon, CD and Lee, RG (1988) Inactivation of biofilm bacteria. *Applied and Environmental Microbiology* 54:2492–2499.

Line, MA (1983) Catalase activity as an indicator of microbial colonization of wood. In: Biodeterioration 5 (Eds. Oxley, TA and Barry, S), John Wiley, New York, pp. 38–43.

McCoy, W, Bryers, JD, Robbins, J and Costerton, JW (1981) Observations in fouling biofilm formation. *Canadian Journal of Microbiology* **27**:910–917.

Meyer-Reil, LA (1978) Autoradiography and epifluorescence microscopy combined for the determination of number and spectrum of actively metabolizing bacteria in natural waters. *Applied and Environmental Microbiology* **36**:506–512.

Mittelman, MW and Geesey, GG (1987) Biological fouling in industrial water systems. Water Micro Associates, San Diego, P.O. Box 28848, San Diego, CA 92128-0848, pp. 269–347.

Nickels, J, Bobbie, RJ, Lott, DF, Maritz, RF, Benson, PH and White, DC (1981) Effect of manual brush cleaning on biomass and community structure of microfouling film formed on aluminium and titanium surfaces exposed to rapidly flowing seawater. *Applied and Environmental Microbiology* **41**:1442–1453.

Nivens, DE, Chambers, JQ and White, DC (1991) Non-destructive monitoring of microbial biofilms at solid–liquid interface using on-line devices. In: Microbially influenced corrosion and biodeterioration (Eds, Dowlings, NJ, Mittelman, MW and Danko, JC), pp. 5.47–5.56.

Obst, U and Holzapfel-Pschorn, A (1988) Enzymatische Tests für die Wasseranalytik. R. Oldenbourg Verlag, München.

Palmgren, R and Nielsen, PH (1996) Accumulation of DNA in the exopolymeric matrix of activated sludge and bacterial cultures. *Water Science Technology* **34**:233–240.

Paulus, W (1993) Microbicides for the protection of materials. Chapman & Hall, London.

Payne, KR (1988) Industrial biocides. Crit. Rep. Appl. Chem. 23, John Wiley, London, New York, pp. 118.

Ridgway, HF and Flemming, H-C (1996) Biofouling of membranes. In: Water treatment membrane processes (Eds, Mallevialle, J, Odendaal, PE and Wiesner, MR), McGraw-Hill, New York, pp. 6.1–6.62.

Roe, FL, Wentland, E, Zelver, N, Warwood, B, Waters, R and Characklis, WG (1994) On-line side-stream monitoring of biofouling. In: Biofouling and biocorrosion in industrial water systems (Eds. Geesey, GG, Lewandowski, Z and Flemming, H-C), Lewis Publ., Ann Arbor, MI, pp. 137–150.

Sand, W (1987) Mikrokalorimetrie—ein modernes Meßverfahren für biologische Fragestellungen. *Forum Mikrobiologie* **6/87**:220–223.

Schaule, G and Flemming, H-C (1997) Pathogenic microorganisms in water system biofilms. Ultrapure Water, April 1997, pp. 21–28.

Schaule, G, Flemming, H-C and Ridgway, HF (1993a) Use of 5-cyano-2,3-ditolyl tetrazolium chloride (CTC) for quantifying planctonic and sessile respiring bacteria in drinking water. *Applied and Environmental Microbiology* **59**:3850–3857.

Schaule, G, Kern, A and Flemming, H-C (1993b) RO treatment of dump trickling water—membrane biofouling. A case history. *Desalination Water Reuse* **3/1**:17–23.

Schmidt, H (1983) Verhinderung organischer Ablagerungen und mikrobielle Kontrolle in industriellen Kühlwassersystemen. *Wasser, Luft u. Betrieb* **27**:12–17.

Schmitt, J and Flemming, H-C (1998) FTIR-spectroscopy in microbial and material analysis. *International Biodeterioration and Biodegradation* **41**:1–11.

Szewzyk, U and Schink, B (1988) Surface colonization by and life cycle of *Pelobacter acidigallici* studied in a continuous-flow microchamber. *Journal of General Microbiology* **134**:183–190.

Tamachkiarowa, A and Flemming, H-C (1996) Glass fiber sensor for biofouling monitoring. *DECHEMA Monographs* **133**:31–36.

Turakhia, MH, Cooksey, KE and Characklis, WG (1983) Influence of a calcium-specific chelant on biofilm removal. *Applied and Environmental Microbiology* **46**:1236–1238.

Wiatr, CL (1990) Controlling industrial slime. European Patent : 0388 115 v. 12.3.90.
Young-Bandala, L and Kajdasz, RJ (1983) A rapid method for monitoring microbial fouling in industrial cooling water systems. Proc.-Int. Water Conf., Eng. Soc. West Pa. 44th, pp. 442–446.
Zips, A, Schaule, G and Flemming, H-C (1990) Ultrasound as a mean for detachment of biofilms. *Biofouling* **2**:323–336.

4 Biofilms in the Food and Beverage Industry

4.1 Problems of Biofilms in the Food and Beverage Industry

JOANNA VERRAN[1] and MARTIN JONES[2]
[1]*Department of Biological Sciences, Manchester Metropolitan University, Chester Street, Manchester M1 5GD, UK*
[2]*Unilever Research, Portsunlight Laboratory, Quarry Road East, Bebington, Wirral L63 3JW, UK*

INTRODUCTION

For the food and beverage industries, the problems associated with biofilms and biofouling are significant. However, in these industries, the deposits or accumulations of microorganisms found upon surfaces do not always fit well within the definitions of these terms. In this section, we address problems, actual and potential, associated with microorganisms immobilized by any means to a surfaces. The substratum may be the raw food material, food contact surfaces, or non-food contact surfaces (floors, walls). The microbial flora may be resident or transient, actively attached or passively retained on the surface, multiplying or not (biostasis). In all cases, there is a 'biotransfer potential' (Hood and Zottola 1995; Wirtanen *et al.* 1996), that is the ability of microorganisms present on equipment surfaces, before and after cleaning procedures, to contaminate products during processing. Unacceptable outcomes for the consumer would be product spoilage or, if pathogenic microorganisms are present, foodborne illness, features which impact upon food quality and safety, respectively. For the producer/food processor, the implications encompass equipment damage, plant shut down, legal proceedings and the resulting effect on reputation and cost.

The literature on microbial contamination of foods and food surfaces is extensive, but biofilm-orientated approaches are more recent and the subject of several reviews (Table 4.1.1). *In vitro* based research focuses on interactions between specific microorganisms, particularly those most recently associated with outbreaks of foodborne illness such as *Listeria monocytogenes* (Blackman and Frank 1996; Kumar and Anand 1998; Zottola 1994), and materials used in various aspects of food processing and production. Many of these papers attempt to determine factors underlying

Industrial Biofouling. Edited by J. Walker, S. Surman, J. Jass
©2000 John Wiley & Sons Ltd

Table 4.1.1. Selected reviews on the structure and distribution of biofilms in the food industry

Arnold (1998) Development of bacterial biofilms during poultry processing
Boulange-Petermann (1996) Processes of bioadhesion on stainless steel surfaces and cleanability: a review with special reference to the food industry
Carpentier and Cerf (1993) A review: biofilms and their consequences, with particular reference to hygiene in the food industry
Carpentier *et al.* (1998) Biofilms on dairy plant surfaces: what's new?
Flint *et al.* (1997a) Biofilms in dairy manufacturing plant—description, current concerns and methods of control
Hood and Zottola (1995) Biofilms in food processing
Kumar and Anand (1998) Significance of microbial biofilms in the food industry: a review
Mattila-Sandholm and Wirtanen (1992) Biofilm formation in the industry: a review
Mittelman (1988) Structure and functional characteristics of bacterial biofilms in fluid processing operations
Notermans S (1994) The significance of biofouling to the food industry
Notermans *et al.* (1991) Contribution of surface attachment to the establishment of microorganisms in food processing plants: a review
Wong (1998) Biofilms in food processing environments
Zottola (1994) Microbial attachment and biofilm formation: a new problem for the food industry?
Zottola and Sasahara (1994) Microbial biofilms in the food processing industry— should they be a concern?

cell attachment, biofilm formation and, most importantly, effective procedures for cleaning and sanitation of surfaces. However, the precision in the description of a controlled model system is often far removed from the complex and variable environment which is being modelled and thus extrapolation is limited. Such limitations are acknowledged in other areas of applied microbiology, where biofilm research has a more established history. We will concentrate on *in situ* findings, where the number of publications are considerably lower than on *in vitro* work.

This section will review the range of cell-substratum interactions in the food and beverage industries, illustrate them with selected case studies and attempt to clarify some of the terminology used. Only detrimental implications and outcomes will be described. An understanding of these interactions should assist the development of appropriate methods for detection and control issues, which are addressed in subsequent parts of this chapter.

MICROBIOLOGY

Diversity of Microbial Contamination

A significant majority of the literature on microbial contamination in the food industry is concerned with bacteria. This is not surprising, because of the association of bacteria with spoilage and diseases and their ubiquity in all potential sources of contamination such as raw materials, equipment surfaces, personnel, pests, the environment, including water air via aerosols and dust, and vehicles for post processing contamination (as packaging) (McEldowney and Fletcher 1990). This will be the main focus of the paper.

Fungi are more often associated with food spoilage than food-borne illness, with the notable exception of aflatoxins, but few papers describe a role for fungi in food-associated biofilms. This is somewhat surprising, given the natural tendency of fungal growth to be associated with surfaces (Jones 1995).

Human/animal viruses are not routinely cultured in food laboratories involved with quality control and hygiene monitoring. Health standards for livestock are monitored elsewhere. Abattoirs are a particular focus for BSE-directed concerns (Anon. 1998a). The likelihood of viruses from food materials being transferred to, and surviving on, industrial food processing surfaces for long enough to pose a biotransfer potential and subsequent infection, may seem remote. However, survival of enteroviruses for up to 60 days on domestic surfaces, has been recorded (Abad *et al.* 1994). There has been little work done on the presence or survival of animal viruses in biofilms or fouling in the food industry. Plant viruses are of more concern to the producer than the processor. Bacterial viruses are of concern in the dairy industry, due to the potential susceptibility of starter cultures. Thus mixed cultures of differing phage sensitivities are used. Bacteriophage have the potential to infect cells within biofilms (Doolittle *et al.* 1996; Hughes *et al.* 1998) and a role in biofilm sloughing and ecology has been mentioned.

Protozoa are usually thought of as aquatic microorganisms, with little relevance to the food industry and their involvement with food borne disease tends to be via water and the faecal contamination route. Recently, the role of protozoa as hosts for pathogenic bacteria has been highlighted and both temporary and permanent associations between amoebae and bacteria may occur within biofilms (Brown and Barker 1999). The *Legionella/Acanthamoeba* association has been well documented and the increased resistance of the bacterial pathogen to chemical inactivation in this state was reported (Barker *et al.* 1992). *Vibrio cholerae* and *Pseudomonas aeruginosa* can also survive in amoebal cysts (Brown and Barker 1999). The potential for protozoa to act as reservoirs in other environments including food manufacturing sites deserves attention.

Attached Microorganisms in the Food and Beverages Industry

Although most procedures for determining microbial load and disinfection susceptibility testing are carried out on suspended cells, the majority of microorganisms in the global environment are not planktonic but sessile

(Coghlan 1996). Thus, the validity of testing procedures could be questioned, since surface testing procedures for disinfection (Holah *et al.* 1998) are not standard practice. Quality control testing on liquids such as mineral waters or milks, justifiably count planktonic cells. On solid foods, where many microorganisms are bound to surfaces, they are still counted in suspension after stomaching, and the number of microorganisms (usually bacteria) present on a fixed weight or food volume is calculated. Too high a microbial load will hasten spoilage, reduce shelf life, indicate a lack of appropriate hygiene or production procedures and potential increase in the risk of pathogens. Spoilage could result from localized microbial activity. However, the distribution of cells on the surface and within the bulk of the food, is not considered when assessing spoilage potential.

Cells attached to inert surfaces, such as processing equipment in the dairy industry (Flint *et al.* 1997a), pipes in the water industry (Verran and Hissett, 1999) or on the walls or caps of bottles (glass or plastic) of mineral water (Jones *et al.* 1999), may provide the inoculum for subsequent product contamination. This may occur directly, if the microorganisms are already on the food or indirectly, to food via cross-contamination from another surface, via a liquid phase passing across the substratum or via aerosol. Attached cells may be detected via surface hygiene testing, which involves monitoring of equipment and environmental surfaces (Hawronskyj and Holah 1997; Holah *et al.* 1989), usually by swabs, which provide some indication of biotransfer potential.

An additional concern is the physiological state of the microorganisms. Dead or injured cells that do not give results from direct culture, might provide anchorage and nutrient for other cells (Flint *et al.* 1997a). 'Viable but non-culturable' cells will not be detected by culture, even though their vitality may be indicated by various stains and microscopic examination (Carpentier and Cerf 1993; Flint *et al.* 1997a). Dormant spores on food and equipment surfaces may germinate if conditions are favourable. What is the significance of these in the food industry? How might they impinge on accepted results of standard testing procedures?

Although viable cells are detected via culture of a specimen, this cannot inform us as to whether or not the cells are growing or are merely present on the original surface. The term biofilm is usually used to describe the result of microbial growth on a surface and later sections describe various inter-pretations of biofilm-like accumulations on food and food associated surfaces.

Microbial Interactions

Current methodologies for recovering organisms from environmental sites undoubtedly greatly underestimate the diversity of species present. This may not matter when the objective is solely to establish the presence or absence of a specific pathogen (e.g. *Salmonella*), but is of concern if we wish to obtain an understanding of the true nature of the microbial communities on food or equipment surfaces. The reason for such studies is more than just academic interest, as the interaction between species, particularly within biofilms, may lead to the expression of properties unexpected from studies on individuals, hence an underestimate of the threat posed to product quality or safety.

Common environmental bacteria such as *Klebsiella* spp., *Enterobacter* spp. or *Pseudomonas* spp. readily form slime which can act as a focus for the attachment and retention of pathogens (Hood and Zottola 1997). Jones and Bradshaw (1997) found that *Klebsiella* could act as an anchor species for *Salmonella*. Similarly, pseudomonads, especially *Ps. fragi*, may be a primary colonizer of surfaces which then entraps *Listeria* (Stoodley *et al.* 1994; Sasahara and Zottola 1993; Bourion and Cerf 1996) or *Salmonella* (Leriche and Carpentier 1995). The presence of a species producing ample exopolymeric slimes may be of particular importance in the colonization of surfaces in flowing conditions (Sasahara and Zottola 1993). Several studies have demonstrated that pathogens may benefit from such associations by a greater resistance to disinfection (Burion and Cerf 1996; de Beer *et al.* 1994; Leriche and Carpentier 1995). Skillman *et al.* (1997, 1998) used strains of *Enterobacter agglomerans* and *Escherichia coli* labelled with the highly fluorescent green fluorescent protein, to examine interactions between mixtures of other enterobacteria isolated from the same sites at a food factory. Dual species biofilms had greater strength of attachment and a higher resistance to disinfection than either species alone (Skillman *et al.* 1997).

Many of the examples cited above involve moderately short-term laboratory experiments. Over a longer time, one might expect the establishment of a resident flora adapted to particular process conditions and the presence of transient microorganisms which become temporarily associated with the surface. Probing such population dynamics is very difficult under factory conditions but at least two studies have shown a persistent resident flora. Dodd *et al.* (1987) used plasmid profiling techniques to show a resident *Staphylococcus aureus* population in poultry processing. More recently, Michiels *et al.* (1997), using random amplification of polymorphic DNA (RAPD), found a persistent strain of a fluorescent pseudomonad on a slicing machine that was responsible for a quality defect in sliced turkey (see page 163). In the latter case, the use of the fingerprinting methods enabled this organism to be tracked even though it was a minor part of the total microbial population. With the availability of molecular techniques, it is anticipated that our ability to dissect natural microbial communities will increase.

Competition between species will also occur and Jeong and Frank (1994) investigated whether a competing flora would eliminate *Listeria* from surfaces. They reported that, although some strains reduced the *Listeria* numbers, none were able to eliminate the pathogen completely. Established natural biofilms may contain a diverse range of microorganisms, including fungi (Jones 1995) and protozoa (Brown and Barker 1999). It is probable that industrial environments will contain equally diverse communities but this is an area which has received little attention to date. The potential to manipulate the surface flora to exclude certain pathogens has been discussed, for example using grazing protozoa (Brown and Barker 1999) or bdellovibrios (Fratamico and Cooke 1996), but this also remains largely unexplored.

BIOFILM DEVELOPMENT

Surface Conditioning

The first stage of biofilm development is often described as the conditioning of inert substrata by organic material derived from the aqueous environment. This phenomenon is particularly well described for dental plaque, where the 'pellicle' is of salivary origin. Organic fouling is a more rapid process than microbial contamination (Carpentier and Cerf 1993; Mettler and Carpentier, 1998). Molecules attach more rapidly than cells suspended in fluid, thus the microorganisms attach to this conditioning film. *In vitro*, a conditioning film is not essential for attachment to be achieved (Flint *et al*. 1997a; Kumar and Anand 1998), but it is difficult to determine what occurs *in vivo*. Very few natural food biofilms have been described or examined in any detail. A pellicle has been demonstrated in water and oral biofilms but these have been studied to a far greater extent and are considerably better characterized than food 'biofilms', where the potential diversity of material for surface conditioning is significant.

Soiling is a term used to describe fouling of food contact surfaces by organic material in the absence of microorganisms (Langeveld *et al*. 1972; Masurovsky and Jordan 1958; Wirtanen *et al*. 1996). Soiled surfaces created *in vitro* (or *in situ*), are rather gross and usually employed to determine the effectiveness of cleaning and sanitation regimes. These are not directly comparable with the rapidly formed, chemically complex (Bower *et al*. 1996), thin, surface conditioning films, created via attraction between organic molecules and substratum. Essentially, soil has an effect on cleanability of surfaces. It also affects the attachment of microorganisms to the surfaces and the strength of their attachment (Flint *et al*. 1997a; Wirtanen *et al*. 1996).

In the food industry, highly active materials such as stainless steel, are immediately chemically conditioned prior to any organic conditioning/ soiling. For steel, the 'stainless' property comes from the tendency of the surface to spontaneously form a 'passivated' protective iron and chromium oxide film in air or water. Microorganisms or organic soil, such as fats, oils, food debris or detergent derived material therefore do not attach to stainless steel but to a hydrophobic (Boulange-Petermann *et al.* 1997) carbon derived surface layer on top of the oxide surface which masks the surface properties of stainless steel. The surface energy is inversely proportional to the thickness of the contaminating carbon layer, which is not eliminated by cleaning (Boulange-Petermann 1996).

The properties of the conditioning film are influenced by those of the underlying substratum (Boulange-Petermann 1996). Hydrophobic polymers and metals cleaned with solvents, favour the adhesion of fatty acids. Hydrophobic protein macromolecules adsorb more readily onto high-energy surfaces. The orientation of contaminating conditioning material on the surface may also be altered by the properties of the substratum, thus presenting different ligands for attachment of micro-organisms. The presence or concentration of conditioning film may enhance or reduce adhesion, as in one study (Flint *et al.* 1997a), a dilute milk soil enhanced bacterial adhesion to a greater extent than the undiluted form. However, this molecular theory for attachment may have little relevance if the surface is uneven, since accumulations of soil and microorganisms are affected by the topography as well as by the chemical properties of the substratum. In a study of soiling over a 10 week period, various floor materials tended to have similar surface free energies, indicating that the initial physicochemical properties of unused surfaces, such as surface free energy, are not maintained, and are therefore not related to later surface fouling (Mettler and Carpentier 1998). Additionally, it is not so much the nature of the forces which contribute to attachment that are of concern, it is their strength, since the ease of removal will determine cleanability (Egington *et al.* 1995; Frank and Chiemelewski 1997), but will also enhance biotransfer potential.

Cleaning consists of two processes: a detergent to solubilize soil (Zottola 1994; Zottola and Sasahara 1994), followed by a sanitizer which kills microorganisms. The detergent activity facilitates the access of the sanitizer to the microorganism. The term soil describes organic material and excludes microorganisms. To define soil as something in the 'wrong place at the wrong time' (Holah 1995), or as a more comprehensive term to describe biofouling including microorganisms, merely adds to a general air of imprecision.

Microbial Attachment

The interaction of microorganisms with a surface may be via liquid, implying some active procedure on the part of the organism to ensure firm attachment and avoid being washed away, or via surface–surface contact that is essentially passive transfer. Factors that influence microbial attachment to surfaces and the mechanisms by which attachment is achieved, are described in detail in other publications (Table 4.1.1). In short, microorganisms become attached to surfaces (Figure 4.1.1) via cell–substratum interaction, with or without a conditioning layer, or via retention by surface defects that provide protection from flow or other shear forces. These attached organisms pose a biotransfer potential.

The strength of attachment is important. If high numbers of attached microorganisms are easily removed by cleaning and cleaning is regular, every 4–24 h with sanitizing, then there should be little problem. However, if a few organism are not easily removed then there is opportunity for multiplication between cleaning cycles and the release of daughter cells (Figure 4.1.2).

The preparation of cells used for adhesion studies varies. Cells may be washed and resuspended in a non-nutrient medium (Vanhaecke *et al.* 1990) to monitor short term attachment in a non-growth environment and in a 'non-specific' manner. Attachment is achieved, even to electropolished stainless steel, within 30 s. Cells may be suspended in broth (Frank and Chmielewski 1997) which would allow the conditioning of the substratum by the broth and, if incubation time were sufficient, would also allow microbial growth rather than attachment alone. In the medical literature there is a confusion between the terms adhesion (no growth) and colonization (biofilm formation?) and should continue to be avoided in the burgeoning food literature. Finally, in some studies, cells are added to

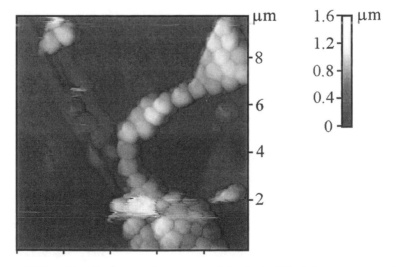

Figure 4.1.1. Atomic force microscopy of bacteria (*Staphylococcus aureus*) on a surface.

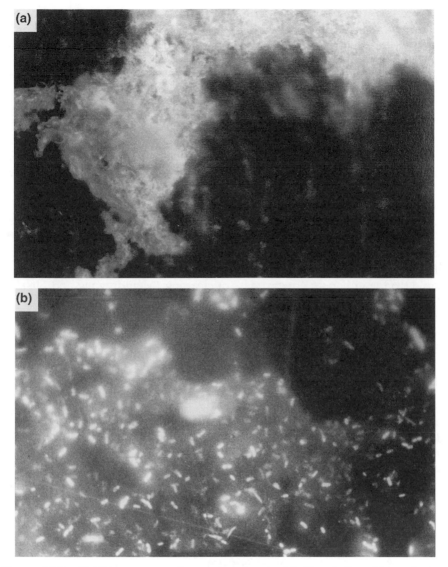

Figure 4.1.2. Epifluorescence microscopy of water biofilms before (2a) and after (2b) a cleaning process. After cleaning, some cells remain attached to the surface, apparently within a diffuse extracellular material.

soil, spread across a test material, dried immediately and exposed to various cleaning agents. This is a study of retention, which describes the resistance of cells to removal (Taylor *et al.* 1998b; Verran and Maryan 1997), rather than adhesion or attachment.

There is a difficulty in terminology concerning 'biofilms'. In general, the term encompasses those cells which have attached to and multiplied on, a surface, often being surrounded by a matrix of extracellular material. In the food industry, the decision as to whether or not to call the accumulation a true biofilm, should be of less concern, than whether or not there is any element of microbial contamination or biofouling on a surface. Hence, biofouling could be defined as occurring when any biological materials (including microorganisms) accumulate on a solid surface (Boulange-Petermann 1996), a definition that would include biofilms. Attached cells, whether in biofilm or not (Carpentier *et al.* 1998), are phenotypically different from their planktonic counterparts, particularly in terms of resistance to antimicrobial agents, and are of additional concern when cleaning and sanitation procedures are being considered.

Even if biofilms are not found on contact surfaces, their presence on non-food contact surfaces (Hood and Zottola 1997; Zottola and Sasahara 1994) could be a problem via aerosols facilitating cross-contamination. Multiplication of microorganisms is enhanced by nutrients, water and protection. Rapid drying enhances effectiveness of cleaning (McEldowney and Fletcher 1990). Thus control methods need to take into account factors which allow the three stages of 'biofilm formation': adsorption, consolidation and colonization (Notermans *et al.* 1991) and the altered properties of the immobilized microorganisms.

Microbial Multiplication

The 'classical' biofilm (Zottola and Sasahara, 1994) morphology is generally taken to be the aquatic model put forward by Costerton and co-workers (1995). Adherent microorganisms multiply to form microcolonies, which enlarge to eventually form vertical 'stacks', separated from one another by water channels. The fluid brings nutrients to the cells. The substratum is irregularly covered, 'streamers' ripple in the flow and parts of the biofilm may detach/slough to enter the fluid phase to potentially seed downstream. Previous perceptions of biofilms comprised an amorphous layer of microorganisms in an extracelluar matrix, rather than this highly hydrated, structured, dynamic entity. However, 'biofilms' appear to take numerous forms in the food industry (Figure 4.1.3), providing considerable diversion from the 'classical' model.

BIOFILMS IN THE FOOD INDUSTRY

Classical Biofilms

The classical biofilm requires the presence of a solid/liquid interface. In the food industry, these may be found in a variety of situations. Static films

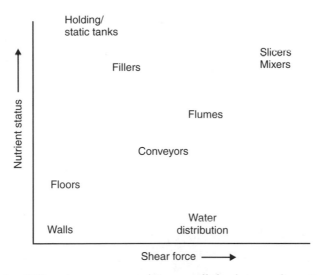

Figure 4.1.3. Different process conditions will lead to a diversity of biofilm structure. Two important factors affecting the extent and type of biofilm formed are the supply of nutrients and hydrodynamic conditions. This figure illustrates just a few of the possible combinations of nutrient supply and shear force that might be found in typical food processing environments.

develop in tank/vats, where the cells may sediment in nutrient medium (Boulange-Petermann 1996). Dynamic films are thicker, found in water distribution systems, drains (Hood and Zottola 1997) and food industry pipelines, where there is liquid flow over the surface (Figure 4.1.3). Other true biofilms have been described on heat exchangers, affecting the efficiency of heat transfer; stirrers, affecting friction; pipelines, causing corrosion or affecting flow; probes, affecting accurate measurement and filters, causing blockages (Kumar and Anand 1998; Matilla-Sandholm and Wirtanen 1992). In general, high flow rates, uneven substratum surface (Figure 4.1.4), infrequent cleaning and a liquid environment appear to facilitate higher microbial counts in biofilms (Bott *et al* 1995; Verran and Hissett 1999; Verran *et al*. 1998).

Process Biofilms

In the food industry, biofilms have rarely been visualized (and then usually via scanning electron microscopy, rather than via less destructive techniques), more often being described in terms of cell density, or substratum coverage. Some have been described as 'process biofilms', which form on surfaces in direct contact with the flowing product and act as a major source of contamination of the product (Flint *et al*. 1997a). The

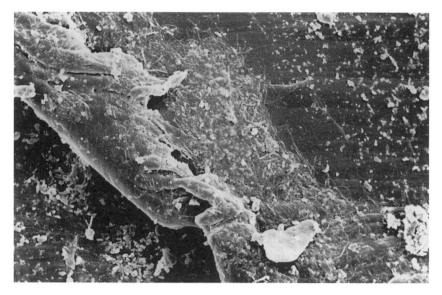

Figure 4.1.4. Scanning electron micrograph demonstrating the effect of flow (left to right) and surface defects on the deposition and accumulation of microorganisms from an aqueous suspension.

dairy biofilm, which forms on the surface of milk processing equipment, would fit this description. It essentially differs from the classical biofilm in the limited range of microorganisms present, in this case to the thermophilic streptococci, the microorganism most likely to be found in the environment. This differs from the diverse heterotrophs in water or the oral commensal flora in plaque, but is comparable with monoculture biofilms described in medical microbiology.

Squashed/Impacted Biofilms

Studies on accumulations of cells in protected environments, where liquid flow is restricted, are few. The continuous depth film fermenter (Ganderton *et al.* 1995) facilitates the accumulation of microorganisms in shallow wells, beneath a moving aqueous environment. The resultant accumulation does not exhibit 'typical' biofilm structure, although the apparatus has enabled studies on the physiology and metabolism of controlled biofilm systems. This type of impacted or 'squashed' biofilm, may be present in teeth fissures and in industry where microorganisms and debris are swept against defects or accumulate in dead ends. Such an accumulation may therefore not show the true biofilm morphology, it may not behave in the tissue-like manner of a true biofilm, nor may it be retained on the surface as firmly as a true biofilm. Although the cells at the base of the accumulation may well be very firmly attached or retained, subsequent cells may have arrived at the site passively, not via active attachment and multiplication (Verran and Hissett 1999).

Biofilms on Foodstuffs

Microbial growth on the surface of foods is well known as part of the ripening processes in cheese where bacterial and fungal coatings develop. Slime formation due to microbial activity is a quality defect found in several foods, including fresh and cooked meats, poultry and vegetables. In addition to surface growth, microorganisms also grow within foods as discrete colonies, termed 'nests', by Katsaras and Leistner (1991). Growth at water–oil interfaces may also occur in emulsions but is not normally regarded as part of the biofilm phenomenon.

Clarification of Terminology

It is hoped that the preceding text has helped to clarify terminology and general principles. The term 'biofouling', when biological fluids are in contact with a solid surface and form a deposit, encompasses biofilms, but is not always the same thing (Notermans 1994). It also describes any soil, with or without attached microorganisms. 'Soil' is best used to describe fouling by organic material without microorganisms. 'Classical' biofilms are rare in the food industry because they are usually controlled by hygienic cleaning. They may be found on surfaces where there is flow of liquid but few have been visualized or characterized in any detail.

To study microbial fouling *in situ*, test materials are placed on site and the resultant microbial attachment is examined (Carpenter and Cerf 1993; Flint *et al.* 1997b; Holah *et al.* 1989; Hood and Zottola 1997; Mettler and Carpentier 1998). In general, coverage is not uniform, cell density is usually less than 10^4 cm^{-2} and the term 'biofilm' is usually used when density exceeds 10^6 cm^{-2} (Flint *et al.* 1997b; Hood and Zottola 1997; Mettler and Carpentier 1998). This low coverage is not surprising, because there is little moisture present and cleaning is regular. However, cleaning procedures themselves may enhance the subsequent fouling of the surface (Boulange-Petermann *et al.* 1997; Mettler and Carpentier 1998), for example by affecting wettability (Boulange-Petermann *et al.* 1993). Wear of materials during usage and cleaning may affect substratum chemistry and topography and subsequent cleanability, a factor which should be considered in plant design strategies and cleaning regimes (Frank and Chmielewski 1997; Matilla-Sandholm and Wirtanen 1992).

The importance of these accumulations in food processing and their prevention and/or control, has increased as have trends towards longer shelf life products; stricter hygiene regulations including an awareness of pathogens; a trend towards longer processing times; novel cleaning and sanitizing regimes; increased automation and complexity and demand for reduction in cost of product, increased quality and fewer additives/ preservatives by a more sophisticated public (Bower et al. 1996; Flint et al. 1997a). Recognition of their presence has necessitated the development of knowledge in terms of associated problems, detection and control.

IMPLICATIONS OF BIOFILMS TO THE FOOD INDUSTRY

Relevance of Laboratory Models

Food processing might be expected to provide ideal conditions for the formation of biofilms, namely a variety of suitable attachment surfaces, ample nutrients and moisture. The raw materials will provide, in many cases, a source of inoculum but water, air and human operators can also be expected to contribute (Holah et al. 1994). One question to be asked is, what is the evidence that biofilms are directly or indirectly responsible for product contamination (Jones 1993), resulting in either product quality defects (tainting, shorter shelf life etc.) or increased risk of pathogen transfer? Additional problems attributable to biofilms might be manifest as increased costs due to equipment fouling, reduced efficiency of heat transfer processes, added cleaning costs and even accelerated corrosion.

Although a considerable number of published reviews claim to deal with the problems caused by biofilms, in general or in specific industries (Table 4.1.1), the majority of the scientific evidence cited relates to model studies under controlled laboratory conditions, invariably using single strains of bacteria. It is very pertinent to question how far these studies should be extrapolated to real processes, and frequently poor experimental design limits further data analysis. A number of factors that influence the conclusions include, the time taken for the biofilm to grow, experimental temperature, nutrient source (often 'broth' with readily available nutrients), lack of competing flora and absence of realistic process conditions (e.g. shear, mechanical abrasion). For example, Blackman and Frank (1996) studied Listeria growth on a variety of surfaces and concluded that Listeria biofilms posed a real threat to food manufacturers. We would not disagree, but would point out there was little growth of Listeria in many of their experimental conditions and in the worst case, growth required 7 days. Similarly, Wirtanen and Mattila-Sandholm (1992) found that a minimum of 48 hours was required to establish biofilms of several organisms in both milk and meat broth model systems. In processes where the cleaning and disinfection frequency is equal to or greater than once a day, the growth of organisms to form biofilms may be of far less significance than microbial entrapment in soil (i.e. biofouling).

Attachment of organisms to surfaces and their biotransfer potential, is greatly influenced by the degree and type of soiling (Wirtanen *et al.* 1996, Sections 3.1 and 3.2). Solids are difficult to handle in laboratory models, hence the tendency for investigators to use liquid media. However, solids and liquids with particulates, are common in food processing. It is possible to use more realistic systems. For example, Brouillard-Delattre *et al.* (1994) used microbial tracers in representative soils and an experimental design that allowed interactions between key factors, such as soil type, surface roughness and time to be fully analysed.

A further difference between laboratory and real systems is that, in many processes, surfaces may not be continuously bathed in liquid. Water availability may be important where surfaces are subjected to cycles of wetting and drying and other physical stresses. Although it may allow biofilm formation it will be limited by the overall ability of the flora to develop. Such regimes are difficult to reproduce in the laboratory, but are essential, if we are to properly evaluate the risks due to biofilm growth. If it is not practical to take more of our investigations out of the laboratory and into the plant, then we must endeavour to discover the conditions affecting surface fouling in the operation of interest and model those parameters as closely as possible.

Topography of Food Contact Surfaces

It is generally acknowledged that an increase in substratum surface roughness will increase microbial attachment on the surface. There would be increased soil/nutrients (Frank and Chmielewski 1997) collected within the surface defects, and increased protection from shear forces would enable retention of the microorganisms within the defects (Korber *et al.* 1997). Thus many experimental papers describing the effect of surface roughness on 'adhesion', by measuring the quantity of attached cells, are essentially describing retention, which is of interest in terms of surface cleanability.

The choice of surface material is of great importance when designing and building equipment and processing lines. Some of the wide range of materials used *in situ* have obvious problems as far as biofilms are concerned, such as gaskets and conveyors. Surfaces that are smooth and in good condition reduce fouling and are easier to clean. Stainless steel is the most common food contact material. It is stable at a variety of temperatures, inert, relatively resistant to corrosion and may be treated mechanically or electrolytically, to obtain surfaces which are easy to clean (Arnold 1998; Boulange-Petermann et al. 1997; Lee 1998; Matilla-Sandholm and Wirtanen 1992). It is used throughout the production chain, from manufacture to storage and food preparation in large scale catering kitchens (Boulange-Petermann 1996). There is a significant amount of literature on its applications in the food industry.

Descriptions of surface roughness/rugosity are calculated via profilo-metry, whereby a stylus traverses the surface profile, perpendicular to the lay of the topography, and produces a tracing of the profile from which measurements may be made (Anon. 1998, 1990). The most commonly used value is the Ra, centre line average, which is the arithmetic average value of the deviation of the trace above and below its centre line. An Ra of below 0.8 μm is generally deemed to be 'hygienic' (Flint et al. 1997a), that is, easy to clean. However, there has been a lack of good correlation between surface roughness measurements and parameters such as cleanability and micro-bial adhesion/retention on surfaces (Boulange-Petermann 1996; Garry et al. 1995; Langeveld et al. 1972; Masurovsky and Jordan 1958; Taylor et al. 1998a; Verran et al. 1991). There are a number of possible reasons for this:

(i) Roughness measurements, particularly Ra, are statistical values describing a surface characteristic. Ra is only of use if there is a regular topography as for polished and brushed stainless steel surfaces, but not for corroded or pickled surface (Boulange-Petermann et al. 1997) or for surfaces which exhibits random scratching or pitting as signs of wear.

(ii) The calculated Ra value is related to the size of the profilometric stylus probe (Anon. 1988b; Taylor et al. 1998a), but this relationship has not been explored. Atomic force microscopy (AFM) has been suggested (Boulange-Petermann 1996; Boulange-Petermann et al. 1997) as a means for measuring roughness, since the probe is of nanometer dimensions. Different considerations are necessary for floors, an obvious reservoir of contamination (Mettler and Carpentier 1998; Notermans et al. 1991), where anti-slip properties demand roughness on a macro scale, whilst cleanability requires smoothness on a micro scale.

(iii) The sampling length/area of substratum examined profilometrically may vary considerably. This feature is of particular concern if there are random defects on the surface, as might be expected during use as surfaces exhibit wear. These might have a significant effect on Ra measurement if included or excluded from examination. Yet without examination of worn or used surfaces, 'typical' wear cannot be determined. Furthermore, if wear occurs on a micro scale, then only the AFM would be able to measure and visualize the surface topography. We have used dental impression materials to sample used surfaces (Rowe *et al.* 1999) and visualised their wear over time (Figure 4.1.5) using AFM, with a view to recreating typical worn surfaces *in vitro*, in order to test their fouling and cleaning potential.

(iv) The type and degree of roughness will affect the retention of cells and soil on a surface but is not addressed by the Ra measurement (Garry *et al.* 1995; Milledge and Jowitt 1980; Taylor *et al.* 1998a, b; Verran *et al.* 1991; Verran and Maryan 1997).

(v) Other features which affect attachment to the surface include wettability, determined by hydrophobicity (Boulange-Petermann *et al.* 1997), which in turn is affected by surface roughness and electrostatic charge density (Zottola and Sasahara 1994).

One might postulate, rather optimistically, that there is a threshold for roughness (type and degree) below which no fouling occurs (Freeman *et al.* 1990). A hygienic surface would need to resist wear and retain hygiene. By taking impressions of work surfaces (Notermans *et al.* 1991; Rowe *et al.* 1999) over time and characterizing surfaces with AFM, it is now possible to investigate surface roughness and cleanability and monitor wear on the appropriate scale.

Biofilms in the Food Industry

The meat and poultry industries have provided us with some clear examples of problems that can arise from biofilms. These industries also provide several studies where the value of starting research in the meat plant or abattoir prior to subsequent laboratory investigations, is evident.

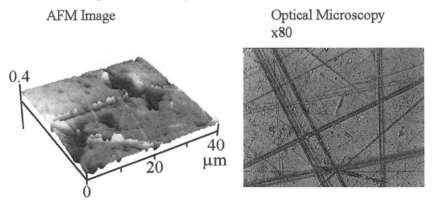

AFM Image

Optical Microscopy
x80

0.4

40
μm

20

0

Ra = 32 ± 8 nm

Figure 4.1.5. Incident light micrograph (right) and atomic force micrograph (left) of an impression taken from worn plastic surface in a food processing environment.

Raw meat and poultry are generally recognized to constitute a micro-biological hazard and the potential for contamination of the carcass surfaces occurs at an early stage of processing. Many studies have shown that once bacteria attach to the surface of meat or poultry, they are difficult to remove. Subsequent growth on that surface compounds the problem. Attachment of Gram-negative bacteria can be rapid, less than 1 minute (Firstenberg-Eden 1981; Butler et al. 1979; Piette and Idziak 1991; Dickson 1991; Fratamico et al. 1996). Differential attachment to various tissue types has been shown (Fratamico et al. 1996). Attached E. coli can grow on chicken skin to form microcolonies within 24 hours and then confluent growth and slime formation proceeds (Mattila and Frost 1988). Slime and off odours resulting from extensive surface growth will certainly result in spoilage and a product being rejected by the consumer (Jay 1986).

Attempts to remove attached bacteria from meat and poultry surfaces have been extensive but largely unsuccessful. Chlorination of chiller water is widely practised but probably makes only a small impact on surface flora of carcasses. Spraying or dipping in a variety of chemicals including organic acids, peroxide and phosphates have all been advocated (Hwang and Beuchat 1995; Bianchi et al. 1994; Notermans 1994; Zhang and Farber 1996). A certain degree of success was claimed for food grade ortho-phosphates against Salmonella (Giese 1993), but effects of phosphates on the total flora was minimal. Again results differ for both bacterial type and the tissues examined. Fratamico et al. (1996) found that removal of E. coli from adipose beef tissue was more effective than from tenderloin. A partial explanation of these effects may come from the study of Rathgeber and Waldrop (1995) who found that under alkaline conditions, some phosphate mixtures remove a layer of fat from the skin of broiler carcasses and take attached bacteria with them. In practice these authors found that phosphate treatments only conferred a short extension of shelf life. Chemical treat-ments often cause skin discoloration and other quality defects at concentrations less than those required to have significant antimicrobial effects.

Given the difficulties in removing bacteria from meat and poultry surfaces and their ability to grow on those surfaces, it is obviously important that surface contamination should be minimized during processing. There is some disagreement on the degree to which biofilm growth on equipment surfaces contribute to product contamination, especially with pathogens. Van der Marel et al. (1988) agreed with earlier studies by Thomas and McMeekin (1980), that chicken carcasses do become contaminated during processing. However, they concluded that Staphy-lococcus aureus contamination arose from the flocks rather than the equipment. Such issues can only be resolved by using techniques that allow specific strains to be followed through the process. For example,

Dodd and co-workers (Dodd *et al.* 1987; Mead and Dodd 1990) used plasmid profiling to demonstrate that the population of *Staphylococcus aureus* on defeathering machines (pluckers) were endemic to those machines and different from the strains on the incoming birds. The process conditions were thought to select for a specific staphylococci sub population.

The value of molecular typing has been further demonstrated by the recent study of Michiels *et al.* (1997), who used random amplification of polymorphic DNA (RAPD) methods to investigate the populations of fluorescent pseudomonads in a meat plant. Over a three month period, 80 different fluorescent strains were identified by this metabolic finger-printing technique. A resident and persistent strain that was not present in the heterogenous flora of raw meat, was shown to occur on a meat mincer and in samples of turkey meat after mincing. Techniques such as RAPD allow a particular member of a mixed community to be tracked over a long period, even when it is only a minor component of the community. Michiels *et al.* (1997) rightly conclude that it is unwise to extrapolate from studies with pure strains of a single species when we do not understand the factors that can lead to the selection of one strain, over many others, to form a biofilm.

Processing, which cuts, minces or even damages the surface of the meat (e.g. pluckers), leads to a release of readily utilizable nutrients for the microflora (Thomas and McMeekin 1980), and is therefore likely to encourage extensive surface growth and facilitate the exchange of microorganisms between food and surface. Non-food contact surfaces, especially environmental surfaces, in meat plants can develop extensive biofilms (Carpentier and Cerf 1993) but should these be of concern? Certainly, several studies have implicated such sites as reservoirs for pathogens such as *Listeria* (Cox *et al.* 1989). Hood and Zottola (1997) demonstrated adherent Gram-negative bacteria growing on stainless steel chips placed adjacent to food contact surfaces, but the cell numbers were small. In contrast, high numbers of bacteria were found on chips in or near floor drains. Similar low levels were observed on a number of process surfaces by Holah *et al.* (1989). Lindsay *et al.* (1996) also found low microbial counts on metal surfaces in poultry plants but, in contrast, counts/cm^2 on non-metallic surfaces were at least 10 fold higher. The significance of the low microbial levels depends on their biotransfer potential and in the case of floors and drains, their aerosolization during cleaning processes. Aerosolization is thought to be significant in the dairy industry (see below) but more work is needed to show whether aerosolized biofilm bacteria have enhanced survival potential, hence pose greater contamination risks (Jones 1993). It has been assumed that the exopolysaccharides which form the biofilm matrix will contribute to desiccation resistance and survival in aerosols, but their effects appear to be small for Gram-negative bacteria (Ophir and Gutnick 1994).

One of the first food industry sectors to provide detailed studies indicating the problems due to adherent bacteria and biofilms, was the dairy industry. All aspects of production from raw milk through to finished products can have problems. In very general terms *Listeria* is frequently recovered from open plant/environmental sites, whereas thermoduric bacteria and *Bacillus* spores, are associated with the processed product. The adhesive properties of spores their attachment to stainless steel surfaces, and their resistance to typical clean-in-place (CIP) regimes are thought to contribute to the problems seen in some milk production (Andersson *et al.* 1995). The extensive study by Nelson (1990) sought to establish where *Listeria* was likely to be found in different types of dairy plants, and data were obtained from over 8000 samples from 62 sites. Fluid milk plants generally had higher levels of *Listeria*-positive sites than butter or cheese factories but none were free from contamination. The ability of *Listeria* to survive for many hours in aerosolized milk, no doubt facilitates its persistence in the factory environment (Spurlock and Zottola 1991). In processing areas, swabs from conveyor surfaces and floor drains showed frequent contamination, but other wet sites were also of concern (Nelson 1990). Although the *Listeria* was surface attached and presumably had grown *in situ*, this study does not tell us much about the nature of the biofilm present.

Flint *et al.* (1997a, b) reported that biofilms in dairy lines can develop over 12 hours, and unlike other food industry sites, these can be dominated by a single bacterial species, e.g. thermoduric streptococci. Growth on the surfaces of plate heat exchangers is a threat to milk quality (Flint *et al.* 1997a, b; Carpentier *et al.* 1998). Austin and Bergeron (1995) described a variety of sites within dairy lines that contained biofilms. Contaminated gaskets were a problem in both raw and processed milk lines, with Buna-n gaskets exhibiting confluent bacterial growth. In a laboratory study, however, Helke and Wong (1994) described inhibitory properties of the same material, once again highlighting the problems of extrapolation of data from laboratory models to the food plant. However, surface topography (see above) may be a more critical factor than the chemical nature of the surface material, in the degree of soiling or microbial retention and ease of hygienic cleaning (Mosteller and Bishop 1993).

Wong and co-workers (Wong 1998; Somers *et al.* 1994) have provided evidence that the presence of *Lactobacillus curvatus* biofilms can produce quality defects in cheese. These non-starter lactics can form biofilms on equipment surfaces and may not be removed during cleaning. Subsequently they can contaminate the next batch of cheese. Finished product including soft cheeses may pick up contamination from environmental biofilms during storage. Notermans (1994) records problems with *Listeria monocytogenes* contamination from wooden storage racks.

Plant surfaces, like most natural surfaces, will have a flora of micro-organisms including bacteria, yeasts and other fungi growing as a biofilm. Although the numbers of microorganisms on freshly harvested vegetables can be high, these are unlikely to include pathogens unless poor agricultural practice has resulted in sewage or faecal contamination. However pathogens can both survive and grow on the surfaces of fruits and vegetables. The growth of *Listeria* has been well documented. In addition, *Salmonella* has been shown to grow on the surface of tomatoes (Zhuang *et al.* 1995) and *Shigella* will survive, if not grow on, prepared vegetables (Rafii *et al.* 1995) and cut lettuce (Davis *et al.* 1988). Various sprouting seeds have also been shown to be potentially hazardous, with *E. coli* O157 proliferating rapidly on the surface of radish sprouts (Hara-Kudo *et al.* 1997) and mung beans (Andrews *et al.* 1982), and *Salmonella* on alfalfa sprouts (Jaquette *et al.* 1996).

Perhaps more than any other food product, the microbiological quality of fruits and vegetables begins to decline as soon as the produce is harvested and this decline can be greatly accelerated by processes such as cooling, washing, slicing and packing. The problems are well known and the work of Splittstoesser *et al.* (1961a, b) clearly demonstrates the extent to which vegetables being prepared for freezing pick up high levels of contamination from equipment post blanching. Conveyors and flumes which are continuously wet and receiving readily utilizable nutrients from damaged vegetables, provide ideal conditions for the development of biofilm slimes.

Many fruits and vegetables are cooled rapidly after harvesting by use of water (hydrocooling). Although the hydrocooling water can be highly contaminated, this does not invariably lead to increased product contam-ination (Reina 1995). The use of chlorine in cooling waters has only a minor effect on microbial loads on vegetable surfaces. The presence of pectinolytic bacteria can quickly lead to loss of quality. During the early development of retail washed pre-packed vegetables, rapid deterioration of products such as carrots were experienced due to the activity of *Erwinia* and other Gram-negative bacteria. This was associated in some cases with massive biofilm development on washing and grading equipment. The desire to pack product wet to maintain an attractive in-pack appearance, also added to the speedy growth of bacteria on the vegetable surfaces.

On leafy vegetables such as cabbage, bacteria are mainly restricted to the outer leaves at harvest. In another factory study, Garg *et al.* (1990) showed a 2–3 log increase in bacterial counts following shredding or slicing. Washing in chlorinated water did not significantly decrease counts on the leaf surfaces.

Prepared salads can have high counts of, mainly, Gram-negative bacteria such as *Enterobacter*, *Pseudomonas* and *Erwinia* which increase in numbers during chill storage (Brocklehurst *et al.* 1987). With ambient storage or temperature abuse, lactic acid bacteria can grow rapidly (Manvell and Ackland 1986). The potential threat of the growth of anaerobic microorganisms such as *Clostridium botulinum*, if vegetables are packed in modified atmospheres, has also been considered (Solomon *et al.* 1990). We are not aware of any studies on whether *C. botulinum* forms biofilms on produce surfaces. Several studies using vegetable leaf surfaces contaminated with *Listeria* have shown that chemical washes are poorly effective (Brackett 1987; Zhang and Farber 1996). In contrast, Adams *et al.* (1989) found washing could be effective but bacteria could survive in hydrophobic 'pockets' or folds in the leaf surface.

Difficulty in wetting the surface of waxy leaves such as cabbage certainly contributes to problems in decontamination. Further, hypochlorite solutions have been observed to cause the opening of cabbage leaf stomata allowing any wash water or surface bacteria greater access to the leaf interior (MV Jones, unpublished data). A recent study by Seo and Frank (1999) also showed that *E. coli* O157 could penetrate cut lettuce leaf either through the cut edges or the stomata.

CONCLUSIONS

It is clear from both laboratory and field data that surface attached microorganisms, whether simply trapped in surface soil or growing on surfaces as biofilms, can constitute a real threat to both the quality and the safety of food products. It is not easy to quantify that risk or to predict, with any confidence, when or where biofilms will constitute a hazard that requires special attention over and above the current cleaning practices. Amongst the questions that remain to be answered (and clearly the answers will vary for different food processes) are:

1. What are the risks that allow the transfer of pathogens from environmental or non-food contact site biofilms to the product? This aspect has been mainly considered with respect to *Listeria* (Zottola and Sasahara 1994) but applies equally to other pathogens.
2. What are the consequences, if organisms are released from biofilms, for Quality Assurance or Quality Control procedures? Release into product streams may be continuous, via shedding of daughter cells, or sporadic, due to sloughing of large amounts of biofilm material. The former results in low level contamination of all samples in a batch, whereas, the latter results in a rare but heavy contamination of isolated samples. Further aerosolization of the biofilm during

cleaning operations might result in a time period post-cleaning when product lines are particularly vulnerable to contamination.

3. How far can biofilm growth and/or biofouling be prevented, by either improvements in equipment design or modified process operating conditions? As described above, surface roughness and damage can result in increased soil and microbial retention and a potential failure of 'normal' cleaning measures to adequately decontaminate the surface. A relationship between flow rates and surface defects on the effectiveness of disinfection has been reported (Korber *et al.* 1997). However, in general, little attention has been paid to the effects of physical conditions such as flow and shear on the structure of soil, biofilm or the potential opportunities to optimize their removal (Stoodley *et al.* 1999; Verran *et al.* 1998). There is some evidence that increased, counter or pulsed flow of cleaning solutions may be effective at disrupting biofilms stabilized to a particular fluid dynamic regime.

4. What frequency of cleaning is necessary to keep the growth of biofilms under control? Biofilms need time to grow and existing cleaning regimes may be adequate to control this. Existing cleaning systems may be effective in removing biofilm, even though it had not been specifically designed for this purpose. Cleaning and disinfection are costly and non-productive, hence, any additional measures need to be fully justified in terms of quality improvements or increased safety margins.

5. How important are biofilms of non-pathogenic organisms in trapping pathogens, even if transiently, enabling them to survive disinfection and also escape detection?

Finally, a food industry status report (Zottola 1994) noted that there was meagre data about the biological composition or morphology of biofilms from actual food contact surfaces. This has not changed. It is to be hoped that recently available molecular probes and physical techniques, such as the atomic force microscope, will encourage more in-field investigations, and help to resolve some of the problems associated with food industry biofilms.

REFERENCES

Abad, FX, Pinto RM and Bosch, A (1994) Survival of enteric viruses on environmental fomites. *Applied and Environmental Microbiology* **60**:3704–3710.

Adams, MR, Hartley, AD and Cox, LJ (1989) Factors affecting the efficacy of washing procedures used in the production of prepared salads. *Food Microbiology* **6**:69–77.

Andersson, A, Ronner, U and Granum, PE (1995) What problems does the food industry have with the spore forming pathogens *Bacillus cereus* and *Clostridium perfringens*? *International Journal of Food Microbiology* **28**:145–155.

Andrews, WH, Mislevic, PB, Wilson, CR, Bruce, VR, Poelma, PL, Gibson, R, Trucksess, MW and Young, K (1982) Microbial hazards associated with bean sprouting. *Journal of the Association of Official Analytical Chemists* **65**:241–248.

Anon. (1998a) Animal Health 1997. The report of the Chief Veterinary Officer. HMSO, London.

Anon. (1988b) BS 1134 Surface texture: Part 1.

Anon. (1990) BS 1134 Surface texture: Part 2.

Arnold, JW (1998) Development of bacterial biofilms during poultry processing. *Poultry and Avian Biology Reviews* **9**:1–9.

Austin, JW and Bergeron, G (1995) Development of bacterial biofilms within dairy processing lines. *Journal of Dairy Research* **62**:509–519.

Barker, J, Brown, MRW, Collier, PJ, Farrell, I and Gilbert, P (1992) Relationship between *Legionella pneumophila* and *Acanthamoeba polyphaga*: Physiological status and susceptibility to chemical inactivation. *Applied and Environmental Microbiology* **58**:2420–2425.

Bianchi, A, Ricke, SC, Cartwright, AL and Gardner, FA (1994) A peroxidase catalysed chemical dip for the reduction of *Salmonella* on chicken breast. *Journal of Food Protection* **57**:301–304.

Blackman, IC and Frank, JF (1996) Growth of *Listeria monocytogenes* as a biofilm on various food-processing surfaces. *Journal of Food Protection* **59**:827–831.

Bott, TR, Mott, IEC and Santos, R (1995) The effect of surface on the development of biofilms under flowing conditions. In: Adhésion des microorganismes aux surfaces: biofilms-nettoyage-désinfection (Eds. Bellon-Fontaine, M-N and Fourniat, J), Technique et Documentation, Paris, pp. 170–179.

Boulange-Petermann, L, Baroux, B and Bellon-Fontaine, M-N (1993) The influence of metallic surface wettability on bacterial adhesion. *Journal of Adhesion Science and Technology* **7**:221–230.

Boulange-Petermann, L (1996) Processes of bioadhesion on stainless steel surfaces and cleanability: a review with special reference to the food industry. *Biofouling* **10**:275–300.

Boulange-Petermann, L, Rault, J and Bellon-Fontaine, M-N (1997) Adhesion of *Streptococcus thermophilus* to stainless steel with different surface topography and roughness. *Biofouling* **11**:201–206.

Bourion, F and Cerf, O (1996) Disinfection efficacy against pure-culture and mixed-population biofilms of *Listeria innocua* and *Pseudomonas aeruginosa* on stainless steel, Teflon® and rubber. *Sciences des aliments* **16**:151–166.

Bower, CK, McGuire, J and Daeschel, MA (1996) The adhesion and detachment of bacteria and spores on food-contact surfaces. *Trends in Food Science and Technology* **7**:152–157.

Brackett, RE (1987) Antimicrobial effect of chlorine on *Listeria monocytogenes*. *Journal of Food Protection* **50**:999–1003.

Brocklehurst, TF, Zaman-Wong, CM and Lund, BM (1987) A note on the microbiology of retail packs of prepared salad vegetables. *Journal of Applied Bacteriology* **63**:409–415.

Brouillard-Delattre, A, Kobilinsky, A, Cerf, O, Alige, S, Gerlot, G and Herry, JM (1994) Methods for measuring the efficiency of cleaning and disinfection processes on open surfaces. *Lait* **74**:79–88.

Brown, MRW and Barker, J (1999) Unexplored reservoirs of pathogenic bacteria: protozoa and biofilms. *Trends in Microbiology* **7**:46–50.

Butler, JL, Stewart, JC, Vanderzant, C, Carpenter, ZL and Smith, GC (1979) Attachment of microorganisms to pork skin and surfaces of beef and lamb carcasses. *Journal of Food Protection* **42**:401–406.

Carpentier, B and Cerf, O (1993) A review: biofilms and their consequences, with particular reference to hygiene in the food industry. *Journal of Applied Bacteriology* **75**:199–511.

Carpentier, B, Wong, ALC and Cerf, O (1998) Biofilms on dairy plant surfaces: what's new? *Bulletin of the International Dairy Federation* **329**:32–35.

Coghlan, A (1996) Slime City. *New Scientist* **2045**:32–36.

Costerton, JW, Lewandowski, Z, Caldwell, DE, Korber, DR and Lappin-Scott, HM (1995) Microbial biofilms. *Annual Review of Microbiology* **49**:711–745.

Cox, LJ, Kleiss, T, Cordier, JL, Cordellanna, C, Konkel, P, Pedrazzini, C, Beumer, R and Siebenga, A (1989) *Listeria* spp. in food processing, non-food and domestic environments. *Food Microbiology* **6**:49–61.

Davis, H, Taylor, JP, Perdue, JN, Stelma, GN, Humphreys, JM, Rowntree R and Greene, KD (1988) A shigellosis outbreak traced to commercially distributed shredded lettuce. *American Journal of Epidemiology* **128**:1312–1321.

de Beer, D, Srinivasan, R and Stewart, PS (1994) Direct measurement of chlorine penetration into biofilms during disinfection. *Applied and Environmental Microbiology* **60**:4339–4344.

Dickson, JS (1991) Attachment of *Salmonella typhimurium* and *Listeria monocytogenes* to beef tissue. *Food Microbiology* **8**:143–151.

Dodd, CER, Adams, BW, Mead, GC and Waites, WM (1987) Use of plasmid profiles to detect changes in strains of *Staphylococcus aureus* during poultry processing. *Journal of Applied Bacteriology* **63**:417–425.

Doolittle, MM, Cooney, JJ and Caldwell, DE (1996) Tracing the interaction of bacteriophage with bacterial biofilms using fluorescent and chromogenic probes. *Journal of Industrial Microbiology* **16**:331–341.

Egington, PJ, Gibson, H, Holah, J, Handley, PS and Gilbert, P (1995) The influence of substratum properties on the attachment of bacterial cells. *Colloids and Surfaces B: Biointerfaces* **5**:153–159.

Firstenberg-Eden, R (1981) Attachment of bacteria to meat surfaces: a review. *Journal of Food Protection* **44**:602–607.

Flint, SH, Bremer, PJ and Brooks, JD (1997a) Biofilms in dairy manufacturing plant—description, current concerns and methods of control. *Biofouling* **11**:81–97.

Flint, SH. Brooks, J, van den Elzen, H and Bremer, P (1997b) Biofilms in dairy manufacturing plant—a threat to product quality. *The Food Technologist* **27**:61–64.

Frank, JF and Chmielewski, RAN (1997) Effectiveness of sanitation with quaternary ammonium compound of chlorine on stainless steel and other domestic food-preparation surfaces. *Journal of Food Protection* **60**:43–47.

Fratamico, PM and Cooke, PH (1996) Isolation of *Bdellovibrios* that prey on *Escherichia coli*. *Journal of Food Safety* **16**:161–173.

Fratamico, PM, Schultz, FJ, Benedict, RC, Buchanan, RC and Cooke, PH (1996) Factors influencing attachment of *Escherichia coli* O157:H7 to beef tissues and removal using selected sanitizing rinses. *Journal of Food Protection* **59**:453–459.

Freeman, WB, Middis, J and Muller-Steinhagen (1990) Influence of augmented surfaces and of surface finish on particulate fouling in double pipe heat exchangers. *Chemical Engineering Processes* **27**:1–11.

Ganderton, L, Kinniment, S and Wimpenny, J (1995) Development and organisation of a steady state, nine-membered community biofilm. In: The life and death of Biofilm (Eds. Wimpenny, J, Handley, P, Gilbert, P and Lappin-Scott, H), Bioline, Cardiff, pp. 37–43.

Garg, N, Churney, JJ and Splittstoesser, DF (1990) Effect of processing conditions on the microflora of fresh cut vegetables. *Journal of Food Protection* **53**:701–703.

Garry, P, Andersen, T, Vendeuvre, JL and Bellon-Fontaine, M-N (1995) Influence de la rugosité de surfaces en polyuréthane sur l'adhésion de *Bacillus subtilis* et *Bacillus cereus*. In: Adhesion des microorganismes aux surfaces: biofilms-nettoyage-desinfection (Eds. Bellon-Fontaine, M-N and Fourniat, J), Technique et Documentation, Paris, pp. 21–30.

Giese, J (1993) *Salmonella* reduction process receives approval. *Food Technology* **46**:110.

Hawronskyj, J-M and Holah, JT (1997) ATP: universal hygiene monitor. *Trends in Food Science and Technology* **8**:79–84.

Hara-Kudo, Y, Konuma, H, Iwaki, M, Kasuga, S, Sugita-Konishi, Y, Ito, Y and Kumagai, S (1997) Potential hazard of radish sprouts as a vehicle for *Escherichia coli* O157:H7. *Journal of Food Protection* **60**:1125–1127.

Helke, DM and Wong, ACL (1994) Survival and growth characteristics of *Listeria monocytogenes* and *Salmonella typhimurium* on stainless steel and Buna-N rubber. *Journal of Food Protection* **57**:963–968.

Holah, JT, Bloomfield, SF, Walker, AJ and Spenceley, H (1994) Control of biofilms in the food industry. In: Bacterial biofilms and their control in medicine and industry (Eds. Wimpenny, J, Nichols, W, Stickler, D and Lappin-Scott, H), Bioline, Cardiff, pp. 163–168.

Holah, JT (1995) Disinfection of food production areas. *Reviews in Science and Technology Office International des Epizootics* **14**:343–363.

Holah, JT, Betts, RP and Thorpe, RH (1989) The use of epifluorescence microscopy to determine surface hygiene. *International Biodeterioration* **25**:147–153.

Holah, JT, Lavaud, A, Peters, W and Dye, KA (1998) Future techniques for disinfectant efficacy testing. *International Biodeterioration and Biodegradation* **41**:273–279.

Hood, SK and Zottola, EA (1995) Biofilms in food processing. *Food Control* **6**: 9–18.

Hood, SK and Zottola, EA (1997) Isolation and identification of adherent Gram negative microorganisms from four meat processing facilities. *Journal of Food Protection* **60**:1135–1138.

Hughes, KA, Sutherland, IW and Jones, MV (1998) Biofilm susceptibility to bacteriophage attack: the role of phage-borne polysaccharide depolymerase. *Microbiology* **144**:3039–3047.

Hwang, C and Beuchat, LR (1995) Efficacy of selected chemicals for killing pathogenic and spoilage microorganisms on chicken skin. *Journal of Food Protection* **58**:19–23.

Jaquette, CB, Beuchat, LR and Mahon, BE (1996) Efficacy of chlorine and heat treatment in killing *Salmonella stanley* inoculated onto alfalfa seed and growth and survival of the pathogen during sprouting and storage. *Applied and Environmental Microbiology* **62**:2212–2215.

Jay, JM (1986) Modern Food Microbiology. Van Nostrand Reinhold, New York, pp. 117–125.

Jeong, DK and Frank, JF (1994) Growth of *Listeria monocytogenes* at 10 C in biofilms with microorganisms isolated from meat and dairy processing environments. *Journal of Food Protection* **57**:567–586.

Jones, CR, Adams, MR, Zhdan, PA and Chamberlain, AHL (1999) The role of surface physiocochemical properties in determining the distribution of the autochthonous microflora in mineral water bottles. *Journal of Applied Microbiology* **86**:917–927.

Jones, K and Bradshaw, SB (1997) Synergism in biofilm formation between *Salmonella typhimurium* and a nitrogen fixing strain of *Klebsiella pneumoniae*. *Journal of Applied Microbiology* **82**:663–668.

Jones, MV (1993) Biofilms and the food industry. In: Bacterial biofilms and their control in medicine and industry (Eds. Wimpenny, J, Nichols, W, Stickler, D and Lappin-Scott, H), Bioline, Cardiff, pp. 113–116.

Jones, MV (1995) Fungal biofilms: eradication of a common problem. In: The Life and Death of Biofilm (Eds. Wimpenny, J, Handley, P, Gilbert, P and Lappin-Scott, H), Bioline, Cardiff, pp. 157–160.

Katsaras, K and Leistner, L (1991) Distribution and development of bacterial colonies in fermented sausages. *Biofouling* 5:115–124.

Korber, DR, Choi, A, Wolfaardt, GM, Ingham, SC and Caldwell, DE (1997) Substratum topography influences susceptibility of *Salmonella enteritidis* biofilms to sodium phosphate. *Applied and Environmental Microbiology* 63:3352–3358.

Kumar, CG and Anand, SK (1998) Significance of microbial biofilms in food industry: a review. *International Journal of Food Microbiology* 42:9–27.

Langeveld, LPM, Bolle, AC and Vegter, JE (1972) The cleanability of stainless steel with different degrees of surface roughness. *Netherland Milk Dairy Journal* 26:149–154.

Lee, J (1998) Bacterial biofilms less likely on electropolished steel. *Agriculture Research* Feb:19–21.

Leriche, V and Carpentier, B (1995) Viable but non-culturable *Salmonella typhimurium* in single- and binary-species biofilms in response to chlorine treatment. *Journal of Food Protection* 58:1186–1191.

Lindsay, D, Geornaras, I and von Holy, A (1996) Biofilms associated with poultry processing equipment. *Microbios* 86:105–116.

Manvell, PM and Ackland, MR (1986) Rapid detection of microbial growth in vegetable salads at chill and abuse temperatures. *Food Microbiology* 3:59–65.

Masruovsky, EB and Jordan, WK (1958) Studies on the relative bacterial cleanability of milk-contact surfaces. *Journal of Dairy Science* 41:1342–1358.

Mattila, T and Frost, AJ (1988) Colonisation of beef and chicken muscle surfaces by *Escherichia coli*. *Food Microbiology* 5:219–230.

Mattila-Sandholm, T and Wirtanen, G (1992) Biofilm formation in the industry: A review. *Food Reviews International* 8:573–603.

McEldowney, S and Fletcher, M (1990) The effect of physical and microbiological factors on food container leakage. *Journal of Applied Bacteriology* 69:190–206.

Mead, GC and Dodd, CER (1990) Incidence origin and significance of staphylococci on processed poultry. *Journal of Applied Bacteriology* 69:81S–91S.

Mettler, E and Carpentier, B (1998) Variations over time of microbial load and physiochemical properties of floor materials after cleaning in food industry premises. *Journal of Food Protection* 61:57–65.

Michiels, CW, Schellekens, M, Soontjens, CCF and Hauben, KJA (1997) Molecular and metabolic typing of resident and transient fluorescent pseudomonad flora from a meat mincer. *Journal of Food Protection* 60:1515–1519.

Milledge, JJ and Jowitt, R (1980) The cleanability of stainless steel used as a food contact surface. *Institute of Food Science and Technology Proceedings* 13:57–62.

Mittelman, M (1988) Structure and functional characteristics of bacteria biofilms in fluid processing operations. *Journal of Dairy Science* 81:2760–2764.

Mosteller, TM and Bishop, JR (1993) Sanitizer efficacy against attached bacteria in a milk biofilm. *Journal of Food Protection* 56:34–41.

Nelson, JH (1990) Where are *Listeria* likely to be found in dairy plants? *Dairy Food and Environmental Sanitation* 10:344–345.

Notermans, S (1994) The significance of biofouling to the food industry. *Food Technology* 48:13–14.

Notermans, S, Dormans, JAMA and Mead, GC (1991) Contribution of surface attachment to the establishment of microorganisms in food processing plants: a review. *Biofouling* **5**:21–36.

Ophir, T and Gutnick, DL (1994) A role for exopolysaccharides in the protection of microorganisms from desiccation. *Applied and Environmental Microbiology* **60**:740–745.

Piette, JPG and Idziak, ES (1991) Role of flagella in the adhesion of *Pseudomonas fluorescens* to tendon slices. *Applied and Environmental Microbiology* **57**:1635–1639.

Rafii, F, Holland, MA, Hill, WE and Cerniglia, CE (1995) Survival of *Shigella flexneri* on vegetables and detection by polymerase chain reaction. *Journal of Food Protection* **58**:727–732.

Rathgeber, BM and Waldrup, AL (1995) Antibacterial activity of a sodium acid pyrophosphate product in chiller water against selected bacteria on broiler carcasses. *Journal of Food Protection* **58**:530–534.

Reina, LD (1995) Microbiological control of cucumber hydrocooling water with chlorine dioxide. *Journal of Food Protection* **58**:541–546.

Rowe, DL, Boyd, RD, Cole, D and Verran, J (1999) Determination of the resolving potential of impression materials using atomic force microscopy. *Journal of Dental Research* **78**:1071.

Sasahara, KL and Zottola, EA (1993) Biofilm formation by *Listeria monocytogenes* utilises a primary colonising microorganism in flowing systems. *Journal of Food Protection* **56**:1022–1028.

Seo, KH and Frank, JF (1999) Attachment of *Escherichia coli* O157:H7 to lettuce leaf surface and bacterial viability in response to chlorine treatment as demonstrated by using confocal scanning laser microscopy. *Journal of Food Protection* **62**:3–9.

Skillman, LC, Sutherland, IW and Jones, MV (1997) Cooperative biofilm formation between two species of enterobacteria. In: Biofilms: Community interactions and control (Eds. Wimpenny, J, Handley, P, Gilbert, P, Lappin-Scott, H and Jones, M), Bioline, Cardiff, pp. 119–122.

Skillman, LC, Sutherland, IW, Jones, MV and Goulsbra, A (1998) Green fluorescent protein as a novel species specific marker in enteric dual-species biofilms. *Microbiology* **144**:2095–2101.

Solomon, HM, Kautter, DA, Lilly, T and Rhodehamel, EJ (1990) Outgrowth of *Clostridium botulinum* in shredded cabbage at room temperature. *Journal of Food Protection* **53**:831–833.

Somers, EB, Wong, ACL and Johnson, M (1994) Biofilm formation by non-starter bacteria and contamination in the dairy plant. *Journal of Dairy Science* **77** (Suppl 1):49.

Splittstoesser, DF, Wettergreen, WP and Pederson, CS (1961a) Control of micro-organisms during preparation of vegetables for freezing I Green bean. *Food Technology* **15**:329–331.

Splittstoesser, DF, Wettergreen, WP and Pederson, CS (1961b) Control of micro-organisms during preparation of vegetables for freezing II Peas and Corn. *Food Technology* **15**:332–334.

Spurlock, AT and Zottola, EA (1991) The survival of *Listeria monocytogenes* in aerosols. *Journal of Food Protection* **54**:910–912, 916.

Stoodley, P, De Beer, D, Lewandowski, Z (1994) Liquid flow in biofilm systems. *Applied and Environmental Microbiology* **60**:2711–2716.

Stoodley, P, Dodds, I, Boyle, JD and Lappin-Scott, HM (1999) Influence of hydrodynamics and nutrients on biofilm structure. *Journal of Applied Microbiology* **85**:60S–69S.

Taylor, RL, Verran, J, Lees, GC and Ward, AJP (1998a) The influence of substratum surface topography on bacterial adhesion to polymethylmethacrylate. *Journal of Materials Science: Materials in Medicine* **9**:17–22.

Taylor, RL, Maryan, C and Verran, J (1998b) Retention of oral microorganisms on cobalt chromium alloy and dental acrylic with different surface finishes. *Journal of Prosthetic Dentistry* **80**:592–597.

Thomas, CJ and McMeekin, TA (1980) Contamination of broiler carcass skin during commercial processing procedures: an electron microscopic study. *Applied and Environmental Microbiology* **40**:133–144.

van der Marel, GM, van Logtestijn, JG and Mossel, DAA (1988) Bacteriological quality of broiler carcasses as affected by in-plant lactic acid decontamination. *International Journal of Food Microbiology* **6**:31–42.

Vanhaecke, E, Remon, J-P, Moors, M, Raes, F, de Rudder, D and van Petegehem, A (1990) Kinetics of *Pseudomonas aeruginosa* adhesion to 304 and 316L stainless steel: role of cell surface hydrophobicity. *Applied and Environmental Microbiology* **56**:788–795.

Verran, J, Lees, GC and Shakespeare, AP (1991) The effect of surface roughness on the adhesion of *Candida albicans* to acrylic. *Biofouling* **3**:183–192.

Verran, J and Maryan, C (1997) Retention of *Candida albicans* on acrylic and silicone of different surface topography. *Journal of Prosthetic Dentistry* **77**:535–539.

Verran, J and Hissett, T (1999) The effect of substratum surface defects upon retention of, and biofilm formation by, microorganisms from potable water. In: Biofilms in the aquatic environment (Eds. Keevil, CW, Holt, D, Dow, C and Godfree, A), RSC, Cambridge, pp. 25–33.

Verran, J, Hissett, T, Beadle, IR and Jones, CL (1998) A simple *in vitro* system for the study of biofilm. *Biofouling* **13**:107–123.

Wirtanen, G and Mattila-Sandholm, T (1992) Effect of growth phase of foodborne biofilms on their resistance to chlorine sanitiser Part II. *Lebensmittel Wissenschaft und Technologie* **25**:50–54.

Wirtanen, G, Husmark, U and Mattila-Sandholm, T (1996) Microbial evaluation of the biotransfer potential from surfaces with *Bacillus* biofilms after rinsing and cleaning procedures in closed food-processing systems. *Journal of Food Protection* **59**:727–733.

Wong, AC (1998) Biofilms in food processing environments. *Journal of Dairy Science* **81**:2765–2770.

Zhang, S and Farber, JM (1996) The effect of various disinfectants against *Listeria monocytogenes* on fresh cut vegetables. *Food Microbiology* **13**:311–321.

Zhuang, RY, Beuchat, LR and Angulo, FJ (1995) Fate of *Salmonella montevideo* on and in raw tomatoes as affected by temperature and treatment with chlorine. *Applied and Environmental Microbiology* **61**:2127–2131.

Zottola, EA (1994) Microbial attachment and biofilm formation: a new problem for the food industry? *Food Technology* **48**:107–114.

Zottola, EA and Sasahara, KC (1994) Microbial biofilms in the food processing industry—should they be a concern? *International Journal of Food Microbiology* **23**:125–148.

4.2 Detection of Biofilms in the Food and Beverage Industry

GUN WIRTANEN, ERNA STORGÅRDS, MARIA SAARELA,
SATU SALO and TIINA MATTILA-SANDHOLM
VTT Biotechnology, Tietotie 2, Espoo, Finland

INTRODUCTION

The hygiene of surfaces, instruments and equipment in the food and beverage industry can have an important effect on the quality of the final product. Microbiological problems in the food industry are often associated with biofilm build-up in pipes, vessels and machinery. Biofilms develop when microbes attached to surfaces secrete extracellular polymers such as polysaccharides and glycoproteins (Costerton *et al.* 1994; Gibson *et al.* 1995b). Biofilm accelerates material deterioration and forms easily on measuring probes and the inner parts of equipment because these are often difficult to clean. Poorly designed sampling valves can destroy the entire process or give rise to misleading information, due to biofilm effects at measuring points. Dead ends, corners, cracks, crevices, gaskets, valves and joints are possible points for biofilm accumulation. Due to their construction, valves are vulnerable to microbial growth and thus, constitute a hygiene risk (Mattila-Sandholm and Wirtanen 1992; Holah and Kearney 1992; Chisti and Moo-Young 1994; Storgårds *et al.* 1999c).

The biofilm formation on surfaces in a laboratory environment is rather difficult to mimic because biofilm is not readily produced in laboratory media rich in nutrients. Bacteria usually do not produce biofilms when nutrients are readily available, and thus the bacteria appear freely suspended (Costerton *et al.* 1987; Anwar *et al.* 1990). Many reviews have been published on the formation and detection of biofilms (Duddridge 1981; Pope and Zintel 1989; Ladd and Costerton 1990; Mattila-Sandholm and Wirtanen 1992; Holah 1992; Gibson *et al.* 1995a, b). Biofilm consists of about 85–96% water, which means that only 2–5% of the total biofilm volume is detectable on dry surfaces (Costerton *et al.* 1981; Duddridge 1981). Methods for studying biofilm formation include microbiological, molecular, biological, microscopical, physical and chemical (Table 4.2.1)

Industrial Biofouling. Edited by J. Walker, S. Surman, J. Jass
©2000 John Wiley & Sons Ltd

Table 4.2.1. Methods for analysing biofilms on food- and technology-related contact surfaces (according to Wirtanen 1995 and Harmsen 1996)

Method	Application, comments
Cultivation	Used for hygiene monitoring Advantage: identification, selective growth media Limitation: sensitivity, time-consuming
Contact plates and dipslides	Used for hygiene monitoring Advantage: identification, selective growth media Limitation: sensitivity, time-consuming
Impedance	Informative in laboratory studies Advantage: more rapid than cultivation, selective growth media Limitation: sensitivity
ATP	Used for hygiene monitoring Advantage: on-site biofilm detectors Limitation: sensitivity, large scatter when analysing multi-layered biofilms due to differing metabolic status of cell layers
Protein residues	Used for hygiene testing Limitation: sensitivity, large scatter
Epifluorescence microscopy	Informative in laboratory studies Advantage: DNA stains and metabolic indicators Limitation: two-dimensional, research purpose
Electron microscopy (SEM, TEM)	Informative in laboratory studies Advantage: structural studies of biofilms Limitation: not for routine use on-site
Scanning confocal laser microscopy (SCLM)	Informative for studies of live biofilms Advantage: can be applied to living biofilms for structural studies Limitation: research purposes only
Molecular biology techniques, e.g. PCR	Informative in laboratory studies Advantage: accurate, sensitive and reproducible method; detection of specific organisms Limitation: not for routine use on-site
Flow cytometry	Informative for studies of microbial densities using various stains Advantages: accurate, sensitive and reproducible method Limitation: detachment of cells; at present mostly used in research due to expensive equipment, small sample volume

(Duddridge 1981; Pope and Zintel 1989; Ladd and Costerton 1990, Mafu *et al.* 1991; Nivens *et al.* 1995; Harmsen 1996; Kostyál, 1998).

In biofilm detection, the planktonic cell counts of the processing fluids should be interpreted with caution because they are not always representative of the sessile organisms (Cloete *et al.* 1989).

Sessile growth of microbes on equipment and other surfaces involved in food production, affects process performance and quality of the final product. The organisms, embedded in a slime matrix of extracellular polymer substances, e.g. polysaccharides and proteins, are well protected against disinfectants and cleaners. Inadequate cleaning and sanitation of surfaces coated with these films represents a source of contamination within the process (Wirtanen 1995). The regrowth of microbial contaminants shortens the interval between cleaning cycles and diminishes process efficiency and product quality (Lelieveld 1985). Practical methods for assessing microbial growth and organic soil on processing surfaces are needed to establish the optimal cleaning frequency of the equipment. To ensure high quality in the end product, reliable detection of micro-organisms with potential detrimental effects is needed throughout the production process (Wirtanen and Mattila-Sandholm 1994; Wirtanen et al. 1995). Quantification of microorganisms from surfaces is difficult due to strong microbial adherence and because the cells grow in layers, forming biofilms. The threshold of detection of adhering microorganisms can vary according to the enumeration technique employed, and some techniques underestimate the number of microorganisms on a surface (Boulangé-Petermann 1996). The problems associated with repeatability and reproducibility in studying biofilm microbes are also well known (Holah et al. 1990).

Reliable monitoring systems that can provide information online, directly and in real time, on microbial growth are required within the process. These monitoring techniques should preferably be adaptable to computer processing. The methods can be based on optical and electrochemical measurements, ion mobility (IM) and infrared (IR) techniques as well as bioluminescence. The successful use of these techniques for on-line monitoring of product quality and cleanliness of surfaces in food-related processes should be based on microbial reference methods. The threshold values for detected numbers of contaminants should be very low (Wirtanen et al. 2000). Advances in molecular biology have resulted in methods with which detection and enumeration of specific organisms on surfaces can be performed (Harmsen 1996; Kostyál 1998).

SAMPLING FROM PROCESS SURFACES FOR MICROBIAL ASSESSMENT

Swabs and Sponges in Surface Sampling

Hygiene monitoring is currently based on conventional cultivation using swabbing or contact plates. In the cultivation of biofilm microorganisms it is important that the sample is detached and mixed properly (Holah *et al.* 1988, 1989). Too strong agitation in detachment of the biofilm from the surface may harm the cells, making them unable to grow on the agar plates, whereas deficient mixing may result in clumps and inaccurate results (Gilbert and Herbert 1987; Holah *et al.* 1988). In the swabbing method the process area of interest is wiped with a cotton- or alginate-tipped swab or sponge to collect possible microorganisms. The microorganisms collected are then released into an appropriate volume of diluent for subsequent cultivation either on agar or in broth. The swab method is easy to use and no specific instrumentation is needed. The advantage of the method is that it provides good access in process areas. The method is also easily adaptable to many detection techniques.

These classical sampling methods, however, suffer from several serious deficiencies (Mittelman 1991). The quantification of cells in the biofilm is difficult to perform because they adhere strongly to the surfaces. It has been reported that cells counted by direct microscopy consistently give results one log unit higher than with the cultivation method (Holah *et al.* 1988; Maukonen *et al.* 2000). This means that techniques based on swabbing provide only limited information on the true status of surface hygiene.

Ultrasonication

Ultrasonics has proved to be an effective tool for the removal of biofilms from surfaces (Zips *et al.* 1990; Bourion and Cerf 1996; Lindsay and von Holy 1997; Salo *et al.* 1999). Cultivation showed that the use of ultrasonics detached about 10 times the number of cells from the surface compared with swabbing (Zips *et al.* 1990). In ultrasonication high-frequency vibration leads to strong formation of very small bubbles that hit the surface at high speed, which in turn causes detachment of surface-bound microorganisms and biofilms. Ultrasonication also produces more reproducible results than the swab technique (Storgårds *et al.* 1999a). The technique has mainly been applied for detaching biofilm bacteria from test slides in bath sonicators. The method cannot be applied as such for routine control of process hygiene.

Detergent Blends for Enhanced Sampling from Surfaces

The use of various detergent and enzyme solutions in cleansing substances is known *per se* in the literature. These publications do not, however, report on the efficiency of the detergents in removing living microbes. Furthermore, the concentrations and temperature of the effective substances used are usually so high that the microbes are damaged as the soil is removed (Tuompo *et al.* 1998). The above-mentioned applications are thus not in themselves suitable for surface hygiene monitoring based on cultivation. At present a method for reliable detachment of microorganisms from process surfaces is still lacking. The detergent-based solutions have been tested using both laboratory and process samples (Wirtanen *et al.* 1997; Salo and Wirtanen 1999). The remaining biofilms were detected by swabbing the surfaces either with swabs moistened in various detergent (alcohol-, surfactant- and enzyme-based) blends or with dry swabs. When dry swabs were used, the detergents were also sprayed on the sampling surface before swabbing. These swabs were placed in a deactivating solution to neutralize the chemicals interfering with the cultivation on agar plates or dipslides (Wirtanen *et al.* 1997). The evaluated test kit comprises a composition designed for the pretreatment of process and equipment surfaces before sampling. It provides a sampling method particularly suited to the monitoring of surface hygiene (Tuompo *et al.* 1998, 1999; Salo and Wirtanen 1999). The kit contains a sampler as well as a detergent-based solution in spray bottles. The cultivation results showed that the use of surface active sampling solution with ultrasonication increased the amount of detachment compared with the use of buffered physiological salt solution with ultrasonication (Salo *et al.* 1999).

TOOLS IN RESEARCH AND INDUSTRIAL USE

Cultivation: Pour and Spread Plates, Contact Agar and Dipslides

Organisms from extreme environments are in many cases difficult to culture, and therefore standard plate counts do not give accurate estimations. Conventional cultivation measures only the number of living cells able to grow on the chosen agar under given circumstances. The plates and slides are usually incubated at 25–30° C for 2–3 days. The agars are either nutrient agars, e.g. which contain tryptose, yeast, glucose and agar–agar, or specific agars in which growth inhibitors e.g. antibiotic or acidic compounds, are added to a nutrient agar base. Contact and culture plates as well as dipslides can be used for flat surfaces; the assumption is made that the microorganisms adhere to the agar surface and begin to multiply on it. The method is easy to use and labour-saving, because it is not necessary to transfer the microbes from the swab or sponge to the cultivation medium. This method, however, is also based on the detachment of surface-bound microorganisms, which is the limiting factor in the swab and sponge method (Cousin 1982; Nortje *et al.* 1989; Merivirta and Uutela 1991; Wirtanen *et al.* 1997; Rahkio and Korkeala 1997; Rahkio, 1998).

Impedance

In the impedance method changes in conductance and capacitance of the samples are measured. Conventional cultivation is based on build-up of biomass, whereas impedance measurement is based on changes in metabolic activity. The impedance changes are dependent on the quantity of ions moving in the liquid: cations moving to the negatively charged electrode and anions to the positively charged electrode (Firstenberg-Eden and Eden 1984). The increase in conductance and capacitance due to the metabolic activity of the microbes leads to a decrease in the impedance. Physical factors affecting the impedance measurement include temperature, composition of the substrate and electrode type. An increase in temperature by $1°$ C increases a change in the capacitance by about 0.9% and a change in the conductance by approximately 1.8% (Firstenberg-Eden and Eden 1984). The temperature must therefore, remain constant so that the impedance measures only the changes due to microbial growth. Measurement of the change in impedance value at suitable time intervals provides an impedance curve and thus enables the detection time of microbial growth in the sample to be identified (Bülte and Reuter 1984; Firstenberg-Eden 1986; Baumgart and Sieker 1993; Salo *et al.* 1996; Flint *et al.* 1997).

The detection time is dependent on the number of microbes in the sample. Handling of the samples, e.g. storage and cooling, can affect them, which means that the detection time for microbial growth may be altered. Impedance measurements must therefore be calibrated with cells treated in the same manner as the samples. Results are achieved more rapidly with impedance measurements than with conventional cultivation (Johnston and Jones 1993). Impedance measurement is used in the food industry to control product quality and to assess the effect of cleaning agents and disinfectants (Holah *et al.* 1990; Mosteller and Bishop 1993; Wirtanen *et al.* 1997). According to Storgårds *et al.* (1999b) impedance measurements revealed that viable bacteria were detected on the gasket materials after cold alkaline–acid cleaning-in-place (CIP) treatment more often with increased age of the gaskets. The cleaning of various materials differed: Viton showed an increased amount of living bacteria after treatments corresponding to 270 CIP cleanings, nitrile–butadiene rubber (NBR) after 380 CIP cleanings and ethylene–propylene terpolymer (EPDM) showed no increase during the study (Storgårds *et al.* 1999b).

ATP and Bioluminescence

The chemical methods used in assessment of biofilm formation are indirect methods based on the utilization or production of specific compounds, e.g. organic carbon, oxygen, polysaccharides and proteins, or on the microbial activity, e.g. living cells and ATP (adenosine 5'-triphosphate) content (Characklis et al. 1982; Ludwicka et al. 1985; Baker et al. 1992; Lundin 1999). The ATP measurement is a luminescence method based on the luciferine– luciferase reaction. The ATP content of the biofilm is proportional to the number of living cells in the biofilm and provides information about their metabolic activity (Gilbert and Herbert 1987; Henriksson and Haikara 1990). Kinetic data obtained for freely suspended cells should not be used to assess immobilized biomass growth, e.g. biofilm. The ATP method is insensitive and therefore not suitable for hygiene measurements in equipment where absolute sterility is needed. According to theoretical limits a reliable ATP value is obtained when there are at least 10^3 bacterial or 10 yeast cells in a sample (Wirtanen 1995).

Bioluminescent bacteria have commonly been used for toxicity testing of environmental and industrial waste samples. Damage to the cellular metabolism caused by toxic compounds can be observed as a change in the light emission (Ribo and Kaiser 1987; Kahru 1993; Lappalainen et al. 1999, 2000). The transformation of luminescence-encoding lux-genes to food-spoilage bacteria or opportunistic pathogens makes it possible to use bioluminescence in monitoring these bacteria e.g. after treatment with cleaning agents (Walker et al. 1993; Flemming et al. 1994).

Microscopy

An important tool in modern biomedical- and technology-related research is fluorescence microscopy. Fluorescence is a type of luminescence in which light is emitted from molecules for a short period of time following the absorption of light (Taylor and Salmon 1989). Fluorescence occurs when an excited electron returns to a lower-energy orbit and emits a photon of light (Angell et al. 1993). Autofluorescence is the fluorescence of naturally occurring molecules in cells (Taylor and Salmon 1989). In Table 4.2.2 various microscopical techniques for studying cell adhesion and biofilm formation on surface materials are listed (Wirtanen 1995; Harmsen 1996; Kostyál 1998).

Scanning (SEM) and transmission (TEM) electron microscopy, has been used to identify biofilm structures (Richards and Turner 1984; Alleman et al. 1985; Costerton et al. 1985; Ladd and Costerton 1990). SEM provides accurate information about the surface materials and the position of biofilm cells but, it does not provide quantitative data for statistical analysis. The electron microscope improves the resolution, but the sample must be dehydrated and embedded in plastic or shadowed with a metal coating (Caldwell et al. 1993). Previously, drying of the SEM preparations destroyed the biofilm structure, however, more recently the use of polyanionic stains, e.g. alcian blue and ruthenium red, has enhanced SEM and TEM

Table 4.2.2. Applications of microscopical techniques in biotechnology and biomedical related research (according to Wirtanen 1995; Harmsen 1996 and Kostyál 1998)

Microscopical techniques*	Application
Electron microscopy:	
1a. immune electron microscopy	1a. identification of microbes
1b. scanning electron microscopy (SEM)	1b. distribution and morphology of microbes on surfaces and in biofilms
1c. transmission electron microscopy (TEM)	1c. distribution and morphology of microbes in biofilms
Epifluorescence microscopy:	
2a. DNA stains: AO, DAPI etc.	2a. distribution of microbes; area coverage
2b. double staining with DNA stains and metabolic indicators: CTC, INT etc.	2b. differentiation of respiring cells in the total amount of cells
2c. fluorescent immuno and genetic probes	2c. identification of microbes; differentiation between types of microbes
Scanning confocal laser microscopy (SCLM):	
3a. fluorescent stains and chemical probes	3a. distribution of microbes, other objects and molecules in biofilms; physiological condition in biofilms; biofilm structures
3b. fluorescent immuno and genetic probes	3b. identification of microbes and their activities in biofilms; ecology of biofilms
Interference reflection microscopy	assessment of adhesive interactions between surfaces and microbes
Differential interference contrast (DIC) microscopy	visualization of individual microbial cells; not for quantitative analysis of biofilm

*The names of the abbreviated chemicals are given in the text.

techniques. Lectins and specific antibodies are also used as stabilizers (Costerton *et al.* 1985).

Microscopic techniques, epifluorescence microscopy as well as SEM and TEM are very informative tools in biofilm research and hygiene studies (Richards and Turner 1984; Lewis *et al.* 1987; Holah *et al.* 1988, 1989; Ladd and Costerton, 1990). One advantage of epifluorescence image analysis is that it allows the study of the cells on the material, rather than cells that have been detached by some method from the surface. In epifluorescence microscopy multilayered biofilms can only be counted two-dimensionally (2-D), which means that the thickness of the biofilm cannot be assessed (Figure 4.2.1). These 2-D imaging techniques have been used to determine the effects of antimicrobial agents on biofilm formation by monitoring

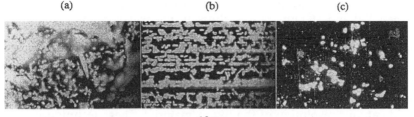

Figure 4.2.1. Epifluorescent microscopy photographs of 5 d old *Pseudomonas fragi* biofilms, which were stained with acridine orange. The biofilms had been grown on various stainless-steel surfaces (AISI 304): (a) glass blasted, (b) lapped and (c) mechanically polished surfaces. The scale marker is equivalent to 10 μm.

attachment rates, detachment rates, growth, viability, morphology and cell area (Caldwell *et al*. 1993). Image analysis is therefore an indispensable tool for obtaining accurate quantitative information from microscopy of the samples. Video technology and increasingly powerful computers facilitate the application of this technique in a wide range of methodologies, including adhesion, surface hygiene, biomass determination and biomedical research (Caldwell *et al*. 1993; Wirtanen 1995).

Scanning confocal laser microscopy (SCLM) is an optical microscopy technique with significant advantages over conventional light microscopy and SEM (Shotton 1989). More accurate information is obtained about the chemical and biological relationships between microorganisms and their microenvironment *in situ* in real time using an SCLM. Attached bacterial cells, microcolonies and biofilms can be effectively studied with SCLM combined with negative fluorescent staining. Advances in software and cryosectioning equipment have improved the applicability of SCLM imaging in biofilm research. Numerous attempts have been made to evaluate the number of living cells in biofilms because it is of interest to count surviving cells after cleaning. Laser microscopy permits optical sectioning of biological materials without optical interference from other focal planes (Caldwell *et al*. 1992a, b, 1993; de Beer *et al*. 1994; Kostyál 1998). The use of image analysis and fluorescent probes in combination with SCLM provides a research tool to analyse changes in biofilm structure and chemistry. SCLM examines the growth and metabolism of living cells in the biofilm; with this technique it is possible to analyse and quantify 3-dimensional (3-D) biofilms and bioaggregates nondestructively, which is not possible with conventional light microscopy (Caldwell *et al*. 1993).

Interference reflection microscopy has been used for investigation of adhesive interactions between the surface and the microorganisms in biofilms. Investigation of microbial growth with lasers at frequencies between 190 and 260 nm causes shifts in the wavelength of the scattered photons and can thus be detected as a resonance Raman spectrum. The quartz-crystal microbalance has proved capable of detecting biomass, but it cannot characterize the constituents in the biofilm (Angell *et al.* 1993). Scanning probe microscopy, including scanning tunnelling electron microscopy, atomic force microscopy and scanning ion-conductance microscopy provide atomic resolution of the living material. These techniques provide only a view of the surfaces of the objects (Caldwell *et al.* 1993; Beech 1996).

Differential interference contrast (DIC) has been used to study crystal formation and dissolution within *Proteus mirabilis* biofilms and also in studies of bacterial biofilms. The software needed in Normarski DIC microscopy is relatively inexpensive compared with the software used in SCLM (McLean *et al.* 1991; Keevil and Walker 1992). Individual micro-organisms have been visualized with acridine orange (AO) and carboxyfluorescein. Normarski DIC microscopy cannot be used for quantitative analysis of the biofilm (Angell *et al.* 1993).

Flow Cytometry

Flow cytometry is a direct optical technique for the measurement of functional and structural properties of individual cells in a cell population. The cells are forced to flow in single file along a rapidly moving fluid stream through a powerful light source. With flow cytometry, single cells can be detected, counted and their viability identified (Storgårds *et al.* 1997). Its advantages include accuracy and speed of analysis (Ueckert *et al.* 1995), sensitivity and reproducibility. Fluorescent probes revealing the physiological properties of the microbes are useful in measuring microbial populations with flow cytometry. Stains such rhodamine 123 (Rh 123), carboxyfluorescein diacetate and Chemchrome B have been used to assess the number of vegetative cells. The light scattered and/or fluoresced by each passing cell is collected and analysed to obtain the number of cells with given properties (Storgårds *et al.* 1997).

Flow cytometry has been used to determine the viability of protozoa, fungi and bacteria. Flow cytometry measures the viability of a statistically significant number of organisms (5000–25 000 cells per sample; Kell *et al.* 1991; Diaper and Edwards 1994a, b; Pore 1994; Mason *et al.* 1995; Lopez-Amoros *et al.* 1997; Storgårds *et al.* 1997). In these respects flow cytometry is suitable for quality control of food raw food materials as well as to inspect hygiene procedures in the process (Ueckert *et al.* 1995). Flow cytometry combined with rRNA-based *in situ* probing has been shown to be a useful tool for rapid and highly automated analysis of the microbial population in biofilms (Amann *et al.* 1990b; Wallner *et al.* 1995). Additional industrial application of flow cytometry in food processing will emerge as soon as the techniques for specifically detecting and analysing food-borne microorganisms have been improved.

Staining Techniques Used in Microscopy and Flow Cytometry

Fluorescence measurements were first applied in the characterization of cultures by Duysens and Amesz (1957), who found that the fluorescence spectra of suspensions of aerobic baker's yeast were similar to that for reduced nicotinamide-adenine dinucleotide phosphate (NAD(P)H). Many different fluorochromes have since been used for the staining of microbes in food samples (Wirtanen 1995). Biofilms and environmental samples have been stained using the following fluorescent dyes: AO (Holah *et al.* 1988, 1989; Davies 1991; Wirtanen and Mattila-Sandholm 1993); Alcian blue (de Beer *et al.* 1994); 5-cyano-2,3-di-p-tolyltetrazolium chloride (CTC) (Schaule *et al.* 1993; Yu and McFeters 1994a, b); 4′,6-diaminidino-2-phenylindole (DAPI) (Porter and Feig 1980); fluorescein isothiocyanate (FITC) (Rogers and Keevil 1992); immunogold (Rogers and Keevil 1992); propidium iodide (de Beer *et al.* 1994); Rh 123 (Yu and McFeters 1994a, b); tetrazolium salts such as nitroblue tetrazolium (NBT), neotetrazolium blue (NT), methylthiazoletetrazolium (MTT) and p-iodonitrotetrazolium violet (INT) (Thom *et al.* 1993) and tritiated uridine incorporation (Yu and McFeters 1994a). The staining with fluorescent dyes in surface hygiene research is rapid and technically easy to perform. Computer-based programs can be used to analyse the area covered with organic material and microbial cells (Holah *et al.* 1988; Caldwell *et al.* 1993; Wirtanen 1995).

In situ detection of microorganisms in food processing can be carried out by microscopic examination of thin sections of food (Dodd 1990; Autio and Mattila-Sandholm 1992) or of surface material (Holah *et al.* 1989; Lawrence *et al.* 1991; Yu *et al.* 1994). The differentiation between viable and dead cells is an important task in studying the microbes present on food contact surfaces (Yu *et al.* 1993; Wirtanen 1995). The use of fluorochrome dyes for detection of viable and dead microbial cells directly on surface materials or in cryosections of the biofilm is possible (Yu *et al.* 1994). Many of the staining techniques described hitherto do not allow differentiation between viable and dead microbial cells, e.g. AO staining is unreliable after heat and irradiation treatments (Betts *et al.* 1988). Rapid identification of viable bacterial spores of *Bacillus* and *Clostridium* spp. was performed with AO staining after germination-like changes had been initiated using performic acid and lysozyme (Sharma and Prasad 1992). In comparisons between viable cell counts determined using the spread plate method and AO staining in epifluorescence microscopy, higher counts were obtained with the microscopy technique (Singh *et al.* 1989). Immunogold labelling of

Legionella pneumophila in biofilms may have advantages over fluorescein isothiocyanite (FITC)-labelling of cells since gold is inherently more stable (Rogers and Keevil 1992). Results in the detection of viable *Listeria monocytogenes* cells showed that CTC staining slightly underestimated and INT overestimated the viability compared with conventional cultivation. CTC, however, proved to be a sensitive indicator of uninjured cells (Schaule *et al.* 1993; Bovill *et al.* 1994; Yu and McFeters 1994a). A treatment with Tris-ethylenediamine tetraacetic acid (Tris-EDTA), which permeabilized the outer membrane, permits good staining with Rh 123. Viable, nonviable and non-viable but resuscitable cells of both Gram-negative and Gram-positive bacteria can rapidly be assessed by the extent of Rh 123 accumulation in the cells (Kaprelyants and Kell 1992).

In some cases the food particles, debris and slime in fluorescent-stained samples can mask the fluorescence of microorganisms. In addition auto-fluorescence can also cause problems (Alcock *et al.* 1990; Maukonen *et al.* 2000). The use of counterstains can be applied to overcome this problem (Sigsgaard 1989; Autio and Mattila-Sandholm 1992; Dufour and Colon 1992). Counterstains, e.g. DAPI–CTC, allow enumeration of active and total bacterial populations in environmental samples within the same prepara-tion (Rodriguez *et al.* 1992). CTC is a soluble redox indicator that can be reduced by respiring bacteria to fluorescent CTC-formazan (Yu and McFeters 1994a, b). CTC in combination with the DNA stain DAPI has been used to differentiate between respiring and nonrespiring cells in biofilms (Stewart *et al.* 1994). Epifluorescence micrographs of frozen biofilm cross-sections clearly revealed gradients of respiratory activity within biofilms (Huang *et al.* 1995). It is also important to use staining techniques that can differentiate stained microorganisms from the background. LIVE/ DEAD BacLight Viability Kit (LIVE/DEAD) provides a 2-colour fluores-cence assay of bacterial viability that has proven applicable for a wide array of bacterial genera including both Gram-negative and Gram-positive bacterial species. All the tested bacterial species have shown good correlation between the results obtained with the LIVE/DEAD and standard plate counts (Maukonen *et al.* 2000). The stains in the LIVE/ DEAD kit are a proprietary mixture of nucleic acid stains that differ both in their spectral characteristics and in their ability to penetrate healthy bacterial cell membranes. LIVE/DEAD staining is, however, not suitable for biofilms attached to the surface, due to interference of biofilm matrix polysaccharides and slime with the stain (Maukonen *et al.* 2000). CTC stained a fluorescent-red dot in live cells, while DAPI stained all the cells fluorescent-green, which made distinguishing easy. CTC–DAPI staining in which CTC was fixed with formalin was suitable for staining of both swabbed biofilm bacteria and biofilms attached to the surface. LIVE/DEAD samples should be analysed immediately, whereas the samples stained with CTC–DAPI can be stored several weeks. CTC–DAPI therefore offers a better tool for biofilm investigation (Bredholt *et al.* 1999; Maukonen *et al.* 2000). Despite these studies there is still a need to develop reliable, simple staining techniques for practical hygiene assessment.

Molecular Techniques

The polymerase chain reaction (PCR) is a method in which thermostable DNA polymerase is used to exponentially amplify a target DNA sequence defined by two oligonucleotide primers (short segments of synthetic DNA). The amplified DNA fragment can be visualized by agarose gel electrophoresis, which also allows the size determination of the PCR product, or by hybridizing the PCR product with a labelled probe. Combining PCR with a hybridization step improves the sensitivity and specificity of the assay. PCR-detection of specific bacteria in a sample material with an abundant microflora (e.g. environmental samples) sets high demands on the specificity of the PCR method applied. In addition, many types of sample materials (e.g. foods) contain factors which can either totally inhibit the amplification reaction or cause partial inhibition leading to a non-exponential amplification of the target DNA (Scheu *et al.* 1998). Inhibition may be avoided or reduced by pre-PCR sample manipulations such as dilution of the sample material, extraction of the DNA from the sample or by harvesting the bacterial cells from the sample e.g. by using immuno-magnetic beads coated with specific monoclonal antibodies. However, even partial inhibition of the PCR reaction inevitably leads to reduced sensitivity and excludes the possibility of performing quantitative PCR. To minimize the risk of obtaining false-negative amplification results suitable external standards, which are coamplified together with the target DNA in the PCR reaction, should be used (Reischl and Kochanowski 1995). Sensitivity of the PCR assay can be improved by short enrichment culture treatment of the sample prior to PCR (Agersborg *et al.* 1997; Wang *et al.* 1997), but this also precludes attempts to quantitate the number of target organisms in the sample. Thus amplification of target DNA sequences from sample material containing inhibitory factors for PCR can provide information on the presence, but not on the numbers and usually not on the viability of target organisms. Several variations of the PCR technique have been applied for the detection of food pathogens and contaminants. These include techniques such as multiplex PCR (several targets are amplified in a single PCR reaction, Bhaduri and Cottrell 1998), nested PCR (same target is amplified using first 'outer' and then 'inner' primers, Mäntynen *et al.* 1997), PCR combined to a hybridization step (Deng *et al.* 1996), reverse transcription PCR (RT-PCR; initial target for PCR amplification in mRNA, which allows the detection of viable bacteria, Klein and Juneja 1997) and immunomagnetic PCR (IM–PCR; PCR is preceded by immunomagnetic separation of bacterial cells, Docherty *et al.* 1996). Recently, a technique called *in situ* PCR, where PCR amplification takes place within bacterial cell, was applied in studies on *Salmonella enterica* serovar Typhimurium gene expression (Figure 4.2.2; Tolker-Nielsen *et al.* 1997, 1998). *In situ* PCR may in the future prove applicable in biofilm studies.

Hybridization techniques can be used in microbial detection either alone or combined to a preceding PCR step. During hybridization the labelled probe, usually a denatured DNA fragment, anneals to a denatured target DNA with sequence homology (e.g. genomic DNA or PCR amplification product). The detection of hybrids is based on e.g. radioactive signal, fluorescence or colour reaction, depending on the type of the label. By determining the intensity of the hybridization signal, the number of target organisms can be estimated. During the past years PCR (alone or combined to a hybridization step) has in large extent replaced the previously applied hybridization techniques such as dot–blot or slot–blot hybridization in the detection of bacteria. However, *in situ* hybridization has proved useful in several applications. In this technique the bacterial cell membrane is permeabilized to allow penetration of the labelled oligonucleotide probe and the hybridization then takes place inside the bacterial cell (Amann *et al.* 1990a, b). Probes for *in situ* hybridization are usually labelled with fluorescent stains (e.g. fluorescein or rhodamine) which allow the detection of hybrids with techniques such as epifluorescent microscopy, confocal laser scanning microscope or flow cytometer (Harmsen 1996). *In situ* hybridization has been proven an informative tool in detecting unculturable microorganisms and in identifying bacteria in complex ecosystems, e.g. soil, activated sludge and especially biofilms (Amann *et al.* 1992; Wagner *et al.* 1993; Manz *et al.* 1995; Neef *et al.* 1996; Thomas *et al.* 1997; Santegoeds *et al.* 1998).

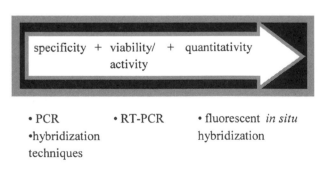

Figure 4.2.2.   Molecular biological detection methods

Physical and Indirect Chemical Methods

The physical methods used in biofilm assessment are either direct or indirect (Busscher et al. 1990). Direct measurements are based on the mass and thickness of the biofilm and indirect methods on friction and heat-transfer assessments. The thickness of the biofilm can be measured with either dry or wet weight (Characklis et al. 1982; Kristensen and Christensen, 1982). Light microscopy has also been used to measure thickness (Bakke and Olsson 1986). It is very important to bear in mind that the biofilm contains 85–96% water (Duddridge 1981). Indirect chemical methods based on measuring the amount of polysaccharides, total organic carbon, total organic nitrogen, proteins, oxygen, organic acids, and total and viable cells can also be used for assessing the thickness of biofilm formations (Characklis et al. 1982).

Viable microbial cells have been enumerated by a peroxyoxalate chemiluminescence assay that determines the concentration of hydrogen peroxide produced in the microorganism's electron transport system (Oh and Cha 1994). The amount of biofilm and its activity were assessed using the ratio of DNA to RNA, the oxygen utilization rate (OUR) or the total amount of nitrogen, muraminic acid or lipopolysaccharides (Pedersen 1982; Gilbert and Herbert 1987; Jørgensen et al. 1992). OUR was found to be in agreement with the ATP measurement of biomass (Jørgensen et al. 1992). A multiple-excitation fluorometric system including four key bacterial fluorophiles (NAD(P)H, pyridoxine, tryptophan and riboflavin) has been used to monitor microbial metabolic processes in bioreactors (Li and Humphrey 1990; Kwong and Rao 1994). The fluorescence emission of tryptophan and NAD(P)H by actively growing biofilms has been measured online using a fluorimeter with a fibre-optic accessory to estimate biofilm mass and metabolic activity (Angell et al. 1993).

Slime formation has been assessed using dyes, e.g. alcian blue (Schadow et al. 1988; Nichols, 1991). Quantitative assessments are based on the measurement of uronic acid or tryptophan (Dall and Herndon 1989) or of slime with colour detectors (Mattila 1989). The activity of microbes in biofilm has been assessed by determining the amount of released carbon dioxide and the amount of glucose used; this assessment can be based on radioactivity counts. Coupons with biofilms were incubated in radioactively labelled glucose and the released radioactive carbon dioxide was measured (Gilbert and Herbert 1987; Ladd and Costerton 1990). The utilization of glucose can also be measured indirectly by using glucose oxidase and peroxidase (Ladd and Costerton 1990). Specific enzymes have been utilized in measuring the phosphatase activity of the biofilm. These include acidic, neutral and alkaline phosphatases, depending on the nature of the microbes forming the biofilm. Phosphatase measurement based on the fluorometric reaction is more sensitive than colorimetric measurement (Collett and Hopton 1987; Brown *et al.* 1988). Furthermore, kits based on assessing protein residues have been developed for measuring the hygiene level in food-processing areas.

FUTURE PROSPECTS FOR ON-LINE MONITORING OF MICROBIAL DEPOSITS

Despite extensive measures carried out to develop reliable, rapid control methods for practical hygiene assessment in the food industry, no such methods are yet available and efforts are still needed for their development. Conventional swabbing procedures for equipment surfaces should be modified e.g. by chemical loosening of the remaining biofilm cells from the surface at the moment of assessment. Practical assessment of microbial films and organic soil on processing surfaces must also be developed for on-line detection, e.g. using Fourier transformation infrared (FTIR) spectrometry. Rapid methods using fluorescent dyes are needed for the detection of living and dead cells left on surfaces after cleaning procedures (Wirtanen *et al.* 1997). Experimental design can make use of several methods, and hopefully new methods will also be developed that could provide accurate data for cleaning regimes in practical hygiene in the food industry (Wirtanen 1995).

The processing equipment should be easily cleaned and "cleanability" testing methods are therefore needed. Development of practical detection methods for research on biofilm build-up is needed to utilize the information on biofilm accumulation in the maintenance of process equipment. The natural luminescence of marine bacteria has been used to assess biofilms in marine systems, for on-line evaluation of biofilm build-up and antifouling coatings (Mittelman *et al.* 1993; Duncan *et al.* 1994). Further efforts are needed to optimize bioluminescent techniques, e.g. using the *Photobacterium leiognathi*, for use in cleanability test procedures. Efforts should focus on adapting the procedures to open and closed systems (Wirtanen 1995).

On-line monitoring techniques used in various industrial fields, e.g. fibre-optic, IR and IM techniques, ion mobility techniques, bioluminescence, microelectrodes as well as heat transfer, could be applied in assessing the quality of raw material and products and for control of microbial and chemical contaminants on processing surfaces. To ensure successful application of these monitoring techniques to food-processing-related industries, studies should be carried out on laboratory, pilot and full-production scales, using both culture and microscopic techniques, as reference methods. The use of measuring techniques in full-scale production must be optimized depending on the threshold values of the method (Wirtanen 2000).

Optical Monitoring Techniques

To meet the particular monitoring requirements for beverage production, a device based on fibre-optic techniques could be developed. A similar method has been used in assessment of deposits on surfaces in environmental science. When a fouling layer accumulates on the front surface, both transmission and reflection of light will change according to the optical properties of the foulant. The thickness of the deposit can be correlated directly for up to several layers of microorganisms (Flemming and Schaule 1996a, b; Tamachikiarowa and Flemming 1996). Measuring the thickness of the deposit can also be used to validate the results obtained with the method applied. The selection of particular light wavelength allows an approximate characterization of the chemical nature of different foulants. The use of cleaners, biodispersants and biocides can thus be optimized, reduced or directed to the most affected areas. New on-line monitoring techniques established for use in food-processing industries to control product quality and microbial as well as chemical contaminants on surfaces should be as user-friendly as possible and the equipment should therefore be modified to suit the process conditions.

Electrochemical Techniques

Electrochemical techniques are widely used for the detection, identification and quantification of organic, organometallic and inorganic compounds at low concentrations. The compound should be extracted from its natural environment before analyses are carried out. The detection of this compound and its concentration is then assessed electrochemically or analytically. Innovations in electrochemistry during the last two decades have shown that the microelectrodes offer many advantages extending the potential application of electrochemistry in many fields. The diminished ohmic drop of microelectrodes enables them to perform as working electrodes in very resistive media previously considered impossible for use in electrochemical experiments (Montenegro 1985, 1994). Another advantage is that they are almost immune to the local hydrodynamics commonly found in solutions, due to the establishment of diffusion layers of lesser dimensions than those of the stagnant hydrodynamic layers imposed at solid/liquid interfaces by thermal convection or by stirring up of flows. The field of microsensors has been progressing rapidly, and it is feasible that microelectrodes could be applied in the detection of microorganism deposits on equipment surfaces (Bento et al. 1998).

Heat-transfer Resistance

The measurement of heat-transfer resistance is based on the fact that a deposit on a surface introduces additional thermal resistance between the source of heat and cold fluids. In industrial heat exchangers, the increase in heat-transfer resistance is usually estimated by measuring the flow rate and the temperatures of the fluids at the inlet and outlet of the heat exchanger. Here, a more accurate technique is likely in the future; three thermocouples will be located in a cross-section of a duct through which the fluid flows, two embedded in the wall of the duct and the third in the fluid. This will enable measurement of the local heat-transfer coefficient between the wall and the bulk fluid (Melo and Pinheiro 1992).

Ion Mobility Techniques

Analysers based on IM techniques have been used in chemical warfare agent detector in factories making explosives and also in automatic leakage identification systems for transportation of chemicals. IM spectrometry is an analytical technique in which analyte molecules are ionized by ion/ molecule reactions in a radioactive ion source and subsequently separated according to different mobilities in a weak electrical field at atmospheric pressure (Paakkanen and Huttunen 1994; Kotiaho et al. 1995). New applications based on IM techniques include measurement of odour and sulphur compounds in the pulp and paper industry and quality control in the food industry. Compounds that are easily detectable include halogenated hydrocarbons, organophosphorus, aldehydes, ketones, halogens and cyanides, nitrogen compounds, most solvents and polyaromatic compounds. Analytical methods based on high-resolution chromatography or methods routinely used in microbiology should be used as reference methods when applications are developed. The IM technique is a promising method for assessment of the quality of raw materials, products and process water (Paakkanen and Huttunen, 1994; Kotiaho et al. 1995).

Infrared Techniques

FTIR has proven useful for the determination of biomass. In FTIR spectroscopy the IR radiation is multiply reflected in an element and at each reflection site a longitudinal radiation wave penetrates into the adjacent environment. This radiation is absorbed by compounds on the surface producing IR absorption bands, which can be used in examination of macromolecules, e.g. proteins, on the surface (Nivens et al. 1993a). Penetration of the IR waves works, however, only in biofilm layers a few cells thick. Monitoring deposit formation in pipes is of great interest in all questions of water transport and contamination. Experiments can be conducted with an FTIR flow-through cell integrated in a bypass pipe holding two IR-transparent windows. The IR beam penetrates the cell after the water is transiently drained out. The signals are collected with a detector and processed into regular transmission FTIR spectra. Initial results have demonstrated that it is possible to recognize microbial deposits and measure the thickness of material being deposited on the surface. It is possible to coat the IR windows with paint or polymers such as polyethylene. The time required for one measurement is dependent simply on the time necessary to drain the cell, and the acquisition of spectra is performed within seconds. The measurement is nondestructive and provides chemical information about the nature of the deposit (Schmitt and Flemming 1996). Bacterial biofilms have also been detected on-line with a quartz-crystal microbalance and IR spectroscopy (Nichols *et al.* 1985; Nivens *et al.* 1993a, b).

SUMMARY OF MAIN METHODS IN HYGIENE MONITORING

Some applications of the main hygiene monitoring methods used both in industry and in research are listed in Table 4.2.3.

Table 4.2.3. Summary of main methods used in hygiene research projects for assessing the cleaning efficiency of process surfaces (according to Wirtanen *et al.* 1997)

Detection method	Application
Conventional cultivation	assesses living bacteria, which are in product samples and released from process surfaces
Image analysis of samples stained with DNA stains e.g. AO, DAPI	assesses biofilm components on the surface (including organic soil, dead and living cells and slime)
Epifluorescence microscopy of samples stained with metabolic indicators, e.g. CTC–DAPI	assesses ratio of living and dead cells in biofilms on surfaces and in liquids
Impedance measurement	assesses the activity of viable and injured bacteria that are able to recover after reactivation in broth, e.g. in disinfectant testing, in products and on process surfaces
ATP measurement	assesses total hygiene
Measurement of protein residues	assesses the amount of protein-based soil

CONCLUSIONS

Methods used for monitoring process hygiene are often based on conventional cultivation using various types of agar plates or ATP. Conventional cultivation requires several days before the result can be obtained, and it measures cells able to form colonies on agar. ATP is used for measuring total hygiene. Detection of deposits built up on equipment surfaces at an early stage enables effective countermeasures and thus results in improvement of the process hygiene. Successful on-line monitoring of microbiological deterioration in the process industry exerts many beneficial impacts of economical and environmental value. On-line monitoring is both economical and environmentally sound when gentle cleaning methods can be used and unnecessary procedures avoided. Many valuable methodological tools used in biofilm research should be developed and applied in industrial monitoring.

REFERENCES

Agersborg, A, Reidun, D and Martinez, I (1997) Sample preparation and DNA extraction procedures for polymerase chain reaction identification of *Listeria monocytogenes* in seafoods. *International Journal of Food Microbiology* **35**:275–280.

Alcock, SC, Chaffey, BJ, Dodd, CER, Morgan, MRA and Waites, WM (1990) Detection of *Bronchotrix thermosphacta* in model systems by antibody linked probes. *Applied Bacteriology* **69**:21.

Alleman, JE, Russell, EK, Lantz Jr, WL and Wukasch, RF (1985) Biofilm cryopreparation for scanning electron microscopy. *Water Research* **19**:1073–1078.

Amann, RI, Krumholz, L and Stahl, DA (1990a) Fluorescent-oligonucleotide probing of whole cells for determinative, phylogenetic, and environmental studies in microbiology. *Journal of Bacteriology* **172**:762–770.

Amann, RI, Binder, BJ, Olson, RJ, Chisholm, SW, Devereux, R and Stahl, DA (1990b) Combination of 16S rRNA-targeted oligonucleotide probes with flow cytometry for analyzing mixed microbial populations. *Applied and Environmental Microbiology* **56**:1919–1925.

Amann, RI, Stromley, J, Devereux, R, Key, R and Stahl, DA (1992) Molecular and microscopic identification of sulfate-reducing bacteria in multispecies biofilms. *Applied and Environmental Microbiology* **58**:614–623.

Angell, P, Arrage, AA, Mittelman, MW and White, DC (1993) On-line, non-destructive biomass determination of bacterial biofilms by fluorometry. *Journal of Microbiolgical Methods* **18**:317–327.

Anwar, H, Dasgupta, MK and Costerton, JW (1990) Testing the susceptibility of bacteria in biofilms to antibacterial agents. *Antimicrobial Agents and Chemotheraphy* **34**:2043–2046.

Autio, K and Mattila-Sandholm, T (1992) Detection of active yeast cells (*Saccharomyces cerevisiae*) in frozen dough sections. *Applied and Environmental Microbiology* **58**:2153–2157.

Baker, JM, Griffiths, MW and Collins-Thompson, DL (1992) Bacterial bioluminescence: applications in food microbiology. *Journal of Food Protection* **55**:62–70.

Bakke, R and Olsson, PQ (1986) Biofilm thickness measurements by light microscopy. *Journal of Microbiolgical Methods* **5**:93–98.

Baumgart, J and Sieker, S (1993) A comparison between Malthus 2000 and BacTrac 4100 in the measurement of *Listeria* growth with a new medium for impedimetric analysis. In: BacTrac 4100 Information 93-3, Sy-Lab, Purkersdorf, Austria, p. 1, 66–75, 85–90 and 105–112.

Beech, IB (1996) Potential use of atomic force microscopy for studying corrosion of metals in the presence of bacterial biofilms—an overview. *International Biodeterioration and Biodegradation* **37**:141–149.

Bento, MF, Thouin, L, Amatore, C and Montenegro, MI (1998) About potential measurements in steady state voltammetry at low electrolyte/analyte concentration ratios. *Journal of Electroanalytical Chemistry* **443**:137–148.

Betts, RP, Farr, L, Bankes, P and Stringer, MF (1988) The detection of irradiated foods using the Direct Epifluorescent Filter Technique. *Journal of Applied Bacteriology* **64**:329–335.

Bhaduri, S and Cottrell, B (1998) A simplified sample preparation method from various foods for PCR detection of pathogenic *Yersinia enterocolitica*: a possible model for other food pathogens. *Molecular and Cellular Probes* **12**:79–83.

Boulangé-Petermann, L (1996) Processes of bioadhesion on stainless steel surfaces and cleanability: a review with special reference to the food industry. *Biofouling* **10**:275–300.

Bourion, F and Cerf, O (1996) Disinfection efficacy against pure-culture and mixed-population biofilms of *Listeria innocua* and *Pseudomonas aeruginosa* on stainless steel, Teflon® and rubber. *Sciences des aliments* **16**:151–166.

Bovill, RA, Shallcross, JA and Mackay, BM (1994) Comparison of the fluorescent redox dye 5-cyano-2,3-ditolyltetrazolium chloride with *p*-iodonitrotetrazolium violet to detect metabolic activity in heat-stressed *Listeria monocytogenes* cells. *Journal of Applied Bacteriology* **77**:353–358.

Bredholt, S, Maukonen, J, Kujanpää, K, Alanko, T, Olofson, U, Husmark U, Sjöberg, A-M and Wirtanen, G (1999) Microbial methods for assessment of cleaning and disinfection of food-processing surfaces cleaned in a low-pressure system. *European Food Research and Technology* **209**:145–152.

Brown, MRW, Allison, DG and Gilbert, P (1988) Resistance of bacterial biofilms to antibiotics: a growth-rate related effect? *Journal of Antimicrobial Chemotheraphy* **22**:777–780.

Busscher, HJ, Bellon-Fontaine, M-N, Mozes, N, Mei, HC, van der Sjollema, J, Cerf, O and Rouxhet, PG (1990) An interlaboratory comparison of physicochemical methods for studying the surface properties of microorganisms. Application to *Streptococcus thermophilus* and *Leuconostoc mesenteroides*. *Journal of Microbiological Methods* **12**:101–115.

Bülte, M and Reuter, G (1984) Impedance measurement as a rapid method for the determination of the microbial contamination of meat surfaces, testing two different instruments. *International Journal of Food Microbiology* **1**:113–125.

Caldwell, DE, Korber, DR and Lawrence, JR (1992a) Confocal laser microscopy and digital image analysis in microbial ecology. In: Advances in Microbial Ecology (Ed. Marshall, KC), Plenum Press, New York, pp. 1–67.

Caldwell, DE, Korber, DR and Lawrence, JR (1992b) Imaging of bacterial cells by fluorescence exclusion using scanning confocal laser microscopy. *Journal of Microbiological Methods* **15**:249–261.

Caldwell, DE, Korber, DR and Lawrence, JR (1993) Analysis of biofilm formation using 2D vs 3D digital imaging. *Journal of Applied Bacteriology* **74**:52S–66S.

Characklis, WG, Trulear, MG, Bryers, JD and Zelver, N (1982) Dynamics of biofilm processess: Methods. *Water Research* **16**:1207–1216.

Chisti, Y and Moo-Young, M (1994) Cleaning-in-place systems for industrial bioreactors: design, validation and operation. *Journal of Industrial Microbiology* **13**:201–207.

Cloete, TE, Smith, F and Steyn, PL (1989) The use of planktonic bacterial populations in open and closed recirculating water cooling systems for the evaluation of biocides. *International Biodeterioration* **25**:115–122.

Collett, RA and Hopton, JW (1987) A method for the determination of enzymatic activity of microbial films on surfaces. In: Industrial Microbiological Testing (Eds. Hopton, JW and Hill, EC), Blackwell Scientific Publications, Oxford, pp. 69–78.

Costerton, JW, Irvin, RT and Cheng, K-J (1981) The bacterial glycocalyx in nature and disease. *Annual Review of Microbiology* **35**:299–324.

Costerton, JW, Marrie, TJ and Cheng, K-J (1985) Phenomena of bacterial adhesion. In: Bacterial Adhesion (Eds. Savage, DC and Fletcher, M), Plenum Press, New York, pp. 3–43.

Costerton, JW, Cheng, K-J, Geesey, GG, Ladd, TI, Nickel, JC, Dasgupta, M and Marrie, TJ (1987) Bacterial biofilms in nature and disease. *Annual Review of Microbiology* **41**:435–464.

Costerton, JW, Lewandowski, Z, de Beer, D, Caldwell, D, Korber, D and James, G (1994) Biofilms, the customized microniche: a minireview. *Journal of Bacteriology* **176**:2137–2142.

Cousin, MA (1982) Evaluation of a test strip used to monitor food processing sanitation. *Journal of Food Protection* **45**:615–619, 623.

Dall, L and Herndon, B (1989) Quantitative assay of glycocalyx produced by viridans group streptococci that cause endocarditis. *Journal of Clinical Microbiology* **27**:2039–2041.

Davies, CM (1991) A comparison of fluorochromes for direct viable counts by image analysis. *Letter of Applied Microbiology* **13**:58–61.

de Beer, D, Stoodley, P, Roe, F and Lewandowski, Z (1994) Effects of biofilm structures on oxygen distribution and mass transport. *Biotechnology and Bioengineering* **43**:1131–1138.

Deng, MY, Cliver, DO, Day, SP and Fratamico, PM (1996) Enterotoxigenic *Escherichia coli* detected in food by PCR and an enzyme-linked oligonucleotide probe. *International Journal of Food Microbiology* **30**:217–229.

Diaper, JP and Edwards, C (1994a) Flow cytometric detection of viable bacteria in compost. *FEMS Microbiology and Ecology* **14**:213–220.

Diaper, JP and Edwards, C (1994b) The use of fluorogenic esters to detect viable bacteria by flow cytometry. *Journal of Applied Bacteriology* **77**:221–228.

Docherty, L, Adams, MR, Patel, P and McFadden, J (1996) The magnetic immunopolymerase chain reaction assay for the detection of *Campylobacter* in milk and poultry. *Letters in Applied Microbiology* **22**:288–292.

Dodd, CER (1990) Detection of sites of microbial growth in food by cryosectioning and light microscopy. *Food Science and Technology Today* **4**:180–182.

Duddridge, JE (1981) Some techniques commonly used in the study of biological films. In: UKAEA Conference in Progress in the Prevention of Fouling in Industrial Plant. UKAEA, Harwell, pp. 54–67.

Dufour, P and Colon, M (1992) The tetrazolium reduction method for assessing the viability of individual bacterial cells in aquatic environments: improvements, performance and applications. *Hydrobiology* **232**:211–218.

Duncan, S, Glover, LA, Killham, K and Prosser, JI (1994) Luminescence-based detection of activity of starved and viable but nonculturable bacteria. *Applied Environmental Microbiology* **60**:1308–1316.

Duysens, LNM and Amesz, J (1957) Fluorescence spectrophotometry of reduced phosphopyridine nucleotide in intact cells in near-ultraviolet and visible region. *Biochimia et Biophysica Acta* **24**:19–26.

Firstenberg-Eden, R and Eden, G (1984) Impedance microbiology. Research Studies Press Ltd., Hertfordshire, UK, pp. 1–121.

Firstenberg-Eden, R (1986) Elecrical impedance for determining microbial quality of foods. In: Foodborne microorganisms and their toxins: Developing methodology (Eds. Pierson, MD and Stern, NJ), Marcel Dekker Inc., New York, pp. 129–144.

Flemming, H-C and Schaule, G (1996a) Biofouling. In: Microbial deterioration of materials (Eds. Sand, W, Heitz, E and Flemming, H-C), Springer Verlag, Heidelberg, pp. 39–54.

Flemming, H-C and Schaule, G (1996b) Measures against biofouling. In: Microbial deterioration of materials (Eds. Sand, W, Heitz, E and Flemming, H-C), Springer Verlag, Heidelberg, pp. 121-139.

Flemming, CA, Lee, H and Trevors, JT (1994) Bioluminescent most-probable-number method to enumerate *lux*-marked *Pseudomonas aeruginosa* UG2Lr in soil. *Applied and Environmental Microbiology* **60**:3458–3461.

Flint, SH, Brooks, JD and Bremer, PJ (1997) Use of the Malthus conductance growth analyser to determine numbers of thermophilic streptococci on stainless steel. *Journal of Applied Microbiology* **83**:335–339.

Gibson, H, Taylor, JH, Hall, KH and Holah, JT (1995a) Biofilms and their detection in the food industry. In: CCFRA R&D Report No. 1 MAFF Project No. 9885, CCFRA, Chipping Campden, pp. 1–88.

Gibson. H, Taylor, JH, Hall, KH and Holah, JT (1995b) Removal of bacterial biofilms. In: CCFRA R&D Report No. 2 MAFF Project No. 9885, CCFRA, Chipping Campden, pp. 1–88.

Gilbert, PD and Herbert, BN (1987) Monitoring microbial fouling in flowing systems using coupons. In: Industrial Microbiological Testing (Eds. Hopton, JW and Hill, EC), Blackwell Scientific Publications, Oxford, pp. 79–98.

Harmsen, H, (1996) Detection, phylogeny and population dynamics of syntrophic propionate-oxidizing bacteria in anaerobic granular sludge, Landbouwuniversiteit Wageningen, Wageningen, the Netherlands, 153 p.

Henriksson, E and Haikara, A (1990) Disinfection of filling halls in breweries. Mallas ja olut, pp. 132–139 (in Finnish).

Holah, JT (1992) Industrial monitoring: hygiene in food processing. In: Biofilms— Science and Technology (Eds. Melo, LF, Bott, TR, Fletcher, M and Capdeville, B), Kluwer Academic Publishers, Dordrecht, pp. 645–659.

Holah, JT and Kearney, LR (1992) Introduction to biofilms in the food industry. In: Biofilms—Science and Technology (Eds. Melo, LF, Bott, TR, Fletcher, M and Capdeville, B), Kluwer Academic Publishers, Dordrecht, pp. 35–41.

Holah, JT. Betts, RP and Thorpe, RH (1988) The use of direct epifluorescent microscopy DEM and the direct epifluorescent filter technique DEFT to assess microbial populations on food contact surfaces. *Journal of Applied Bacteriology* **65**:215–221.

Holah, JT, Betts, RP and Thorpe, RH (1989) The use of epifluorescence microscopy to determine surface hygiene. *International Biodeterioration* **25**:147–154.

Holah, JT, Higgs, C, Robinson, S, Worthington, D and Spenceley, H (1990) A conductance-based surface disinfection test for food hygiene. *Letter of Applied Microbiology* **11**:255–259.

Huang, C-T, Yu, FP, McFeters, GA and Stewart, PS (1995) Nonuniform spatial patterns of respiratory activity within biofilms during disinfection. *Applied and Environmental Microbiology* **61**:2252–2256.

Johnston, MD and Jones, MV (1993) Evaluation of the BacTrac 4100. In: Sy-Lab Information 93-1, Sy-Lab, Purkersdorf, Austria, 24 p.

Jørgensen, PE, Eriksen, T and Jensen, BK (1992) Estimation of viable biomass in wastewater and activated sludge by determination of ATP, oxygen utilization rate and FDA hydrolysis. *Water Research* **26**:1495–1501.

Kahru, A (1993) *In vitro* toxicity testing using marine luminescent bacteria (*Photobacterium phosphoreum*): the Biotox™ test. *ATLA* **21**:210–215.

Kaprelyants, AS and Kell, DB (1992) Rapid assessment of bacterial viability and viability by rhodamine 123 and flow cytometry. *Journal of Applied Bacteriology* **72**:410–422.

Keevil, CW and Walker, JT (1992) Normanski DIC Microscopy and image analysis of biofilms. *Binary* **4**:93–95

Kell, DB, Ryder, HM, Kaprelyants, AS and Westerhoff, HV (1991) Quantifying heterogeneity: flow cytometry of bacterial cultures. *Antonie van Leeuwenhoek* **60**:145–158.

Klein, PG and Juneja, VK (1997) Sensitive detection of viable *Listeria monocytogenes* by reverse transcription-PCR. *Applied and Environmental Microbiology* **63**:4441–4448.

Kostyál, E (1998) Removal of chlorinated organic matter from wastewaters, chlorinated ground, and lake water by nitrifying fluidized-bed biomass, Hakapaino Oy, Helsinki, Finland, 68 p. + 5 app.

Kotiaho, T, Lauritsen, FR, Degn, H and Paakkanen, H (1995) Membrane inlet ion mobility spectrometry for on-line measurement of ethanol in beer and in yeast fermentation. *Analytical Chimia Acta* **309**:317–325.

Kristensen, GH and Christensen, FR (1982) Application of cryo-cut method for measurements of biofilm thickness. *Water Research* **16**:1619–1621.

Kwong, SCW and Rao, G (1994) Metabolic monitoring by using the rate of change of NAD(P)H fluorescence. *Biotechnology and Bioengineering* **44**:453–459.

Ladd, TL and Costerton, JW (1990) Methods for studying biofilm bacteria. *Methods in Microbiology* **22**:285–307.

Lappalainen, J, Loikkanen, S, Havana, M, Karp, M, Sjöberg, A-M and Wirtanen, G (1999) Rapid detection of detergents and disinfectants in food factories. In: 30th R³-Nordic Contamination Control Symposium (Eds. Wirtanen, G, Salo, S & Mikkola, A), VTT Symposium 193. Libella Painopalvelu Oy, Espoo. ISBN 951-38-5268-7, pp. 199–205.

Lappalainen, J, Loikkanen, S, Havana, M, Karp, M, Sjöberg, A-M and Wirtanen, G (2000) Microbial testing methods for detection of residual cleaning agents and disinfectants—Prevention of ATP bioluminescence measurement errors in the food industry. *Journal of Food Proctection* **63**:210–215.

Lawrence, JR, Korber, DR, Hoyle, BD, Costerton, JW and Caldwell, DE (1991) Optical sectioning of microbial biofilms. *Journal of Bacteriology* **173**:6558–6567.

Lelieveld, HLM (1985) Hygienic design and test methods. *Journal of the Society of Dairy Technology* **38**:14–16.

Lewis, SJ, Gilmour, A, Fraser, TW and McCall, RD (1987) Scanning electron microscopy of soiled stainless steel inoculated with single bacterial cells. *International Journal of Food Microbiology* **4**:279–289.

Li, J-K and Humphrey, AE (1990) Use of fluorometry for monitoring and control of a biorector. *Biotechnology and Bioengineering* **37**:1043–1049.

Lindsay, D and von Holy, A (1997) Evaluation of dislodging methods for laboratory-grown bacterial biofilms. *Food Microbiology* 14:383–390.

Lopez-Amoros, R, Castel, S, Comas-Riu, J and Vives-Rego, J (1997) Assessment of *E. coli* and *Salmonella* viability and starvation by confocal laser microscopy and flow cytometry using rhodamine 123, DiBAC4(3), propidium iodide and CTC. *Cytometry* 29:298–305.

Ludwicka, A, Switalski, LM, Lundin, A, Pulverer, G and Wadström, T (1985) Bioluminescent assay for measurement of bacterial attachment to polyethylene. *Journal of Microbiological Methods* 4:169–177.

Lundin, A (1999) ATP detection of biological contamination. In: 30th R^3-Nordic Contamination Control Symposium (Eds. Wirtanen, G, Salo, S & Mikkola, A), VTT Symposium 193. Libella Painopalvelu Oy, Espoo. ISBN 951-38-5268-7, pp. 337–352.

Mafu, AA, Roy, D, Savoie, L and Goulet, J (1991) Bioluminescence assay for estimating the hydrophobic properties of bacteria as revealed by hydrophobic interaction chromatography. *Applied and Environmental Microbiology* 57:1640–1643.

Manz, W, Amann, R, Szewzyk, R, Szewzyk, U, Stenstrom, TA, Hutzler, P and Schleifer, KH (1995) *In situ* identification of legionellaceae using 16S ribosomal-RNA-targeted oligonucleotide probes and confocal laser scanning microscopy. *Microbiology* 141:29–39.

Mason, DJ, Lopez-Armoros, R, Allman, R, Stark, JM and Lloyd, D (1995) The ability of membrane potential dyes and calcafluor white to distinguish between viable and non-viable bacteria. *Journal of Applied Bacteriology* 78:309–315.

Mattila, T (1989) A staining technique to measure capsular polysaccharides of *Staphylococcus aureus* on filter membranes. *Journal of Microbiological Methods* 9:323–331.

Mattila-Sandholm, T and Wirtanen, G (1992) Biofilm formation in the industry: a review. *Food Review International* 8:573–603.

Maukonen, J, Mattila-Sandholm, T and Wirtanen, G (2000) Metabolic indicators for assessing bacterial viability in hygiene sampling using cells in suspensions and swabbed biofilms. *Lebensmittel-Wissenschaft und-Technologie* 33:225–234.

McLean, RJC, Lawrence, JR, Korber, DR and Caldwell, DE (1991) *Proteus mirabilis* biofilm protection against struvite crystal dissolution and its implications in struvite urolithiasis. *Journal of Urology* 146:1130–1142.

Melo, LF and Pinheiro, MM (1992) Biofouling in heat exchangers. In: Biofilms—Science and Technology (Eds. Melo, LF, Bott, TR, Fletcher, M and Capdeville, B), Kluwer Academic Publishers, Dordrecht, pp. 499–509.

Merivirta, L and Uutela, P (1991) Hygiene index—a possible way to improve the hygiene knowledge in slaughterhouses and meat processing plants. *Suomen Eläinlääkärilehti* 96:590–592 (in Finnish).

Mittelman, MW (1991) Bacterial growth and biofouling control in purified water systems. In: Biofouling and biocorrosion in industrial water systems (Eds. Flemming, H-C and Geesey, GG), Springer-Verlag, Berlin Heidelberg, pp. 133–154.

Mittelman, MW, Packard, J, Arrage, AA, Bean, SL, Angell, P and White, DC (1993) Test systems for determining antifouling coating efficacy using on-line detection of bioluminescence and fluorescence in a laminar-flow environment. *Journal of Microbiological Methods* 18:51–60.

Montenegro, MI (1985) The advantages of microelectrodes in the study of electrochemistry. *Portugaliae Electrochimica Acta* 3:165.

Montenegro, MI (1994) Application of microelectrodes in kinetics. In: Research in Chemical Kinetics, Vol. 2 (Eds. Compton, RG and Hancock, G), Elsevier Publishers Ltd, Essex, pp. 1–80.

Mosteller, TM and Bishop, JR (1993) Sanitizer efficacy against attached bacteria in a milk biofilm. *Journal of Food Protection* **56**:34–41.

Mäntynen, V, Niemelä, S, Kaijalainen, S, Pirhonen, T and Lindström, K (1997) MPN-PCR-quantification method for staphylococcal enterotoxin *c1* gene from fresh cheese. *International Journal of Food Microbiology* **36**:135–143.

Neef, A, Zaglauer, A, Meier, H, Amann, R, Lemmer, H and Schleifer, KH (1996) Population analysis in a denitrifying sand filter: Conventional and *in situ* identification of *Paracoccus* spp. in methanol-fed biofilms. *Applied and Environmental Microbiology* **62**: 329–4339.

Nichols, PD, Henson, JM, Guckert, JB, Nivens, DE and White, DC (1985) Fourier transform-infrared spectroscopic methods for microbial ecology: analysis of bacteria, bacterial polymer mixtures and biofilms. *Journal of Microbiological Methods* **4**:79–94.

Nichols, WW (1991) Biofilms, antibiotics and penetration. *Review in Medical Microbiology* **2**:177–181.

Nivens, DE, Chambers, JQ, Anderson, TR, Tunlid, A, Smit, J and White, DC (1993a) Monitoring microbial adhesion and biofilm formation by attenuated total reflection/Fourier transform infrared spectroscopy. *Journal of Microbiological Methods* **17**:199–213.

Nivens, DE, Chambers, JQ, Anderson, TR and White, DC (1993b) Long-term on-line monitoring of microbial biofilms using a quartz crystal microbalance. *Analytical Chemistry* **65**:65–69.

Nivens, DE, Palmer Jr, RJ, and White, DC (1995) Continuous nondestructive monitoring of microbial biofilms: a review of analytical techniques. *Journal of Industrial Microbiology* **15**:263–276.

Nortje, GL, Nel, L, Jordaan, E, Naude, RT, Holzapfel, WH and Grimbeek, RT (1989) A microbiological survey of fresh meat in the supermarket trade. Part 2: Beef retail cuts. *Meat Science* **25**:99–112.

Oh, SK and Cha, SH (1994) Effect of some factors on determination of viable microbial cells by peroxyoxalate chemiluminescence method. *Journal of Applied Bacteriology* **76**:275–281.

Paakkanen, H and Huttunen, J (1994) Gas detector based on ion mobility. *Finnish Air Pollution Prevention News* **5–6**:20–23.

Pedersen, K (1982) Method for studying microbial biofilms in flowing-water systems. *Applied and Environmental Microbiology* **43**:6–13.

Pope, DH and Zintel, TP (1989) Methods for investigating underdeposit microbiologically influenced corrosion. *Material Performance* **2811**:46–51.

Pore, RS (1994) Antibiotic susceptibility testing by flow cytometry. *Journal of Antimicrobial Chemotheraphy* **34**:613–627.

Porter, KG and Feig, YS (1980) The use of DAPI for identifying and counting aquatic microflora. *Limnology and Oceanography* **25**:943–948.

Rahkio, TM and Korkeala, HJ (1997) Use of Hygicult-tpc® in slaughterhouse hygiene control. *Acta Veterinary Scandinavia* **38**:331–338.

Rahkio, M (1998) Studies on factors affecting slaughterhouse hygiene. Yliopistopaino, Helsinki, Finland, 52 p. + 5 app.

Reischl, U and Kochanowski, B (1995) Quantitative PCR—a survey of the present technology. *Molecular Biotechnology* **3**:55–71.

Ribo, JM and Kaiser, KLE (1987) *Photobacterium phosphoreum* toxicity bioassay. I. Test procedures and applications. *Toxicity Assessment* **2**:305–323.

Richards, SR and Turner, RJ (1984) A comparative study of techniques for the examination of biofilms by scanning electron microscopy. *Water Research* **18**:767–773.

Rodriguez, GG, Phipps, D, Ishiguro, K and Ridgway, HF (1992) Use of fluorescent redox probe for direct visualization of actively respiring bacteria. *Applied and Environmental Microbiology* **58**:1801–1808.

Rogers, J and Keevil, CW (1992) Immunogold and fluorescein immunolabelling of *Legionella pneumophila* within an aquatic biofilm visualized by using episcopic differential interference contrast microscopy. *Applied and Environmental Microbiology* **58**:2326–2330.

Salo, S, Wirtanen, G and Mattila-Sandholm, T (1996) Impedimetric assessment of foodborne biofilms on stainless steel and teflon surfaces. In: Future Prospects of Biofouling and Biocides, VTT Symposium 165 (Eds. Wirtanen, G, Raaska, L, Salkinoja-Salonen, M and Mattila-Sandholm, T), VTT Offsetpaino, Espoo. ISBN 951-38-4556-7, p. 65.

Salo, S and Wirtanen, G (1999) Detergent based blends for swabbing and dipslides in improved surface sampling. In Biofilms: The Good, the Bad and the Ugly (Eds. Gilbert, P, Brading M, Walker, J and Wimpenny, J), BBC4, Bioline, Cardiff, ISBN 0-9520432-6-2, pp. 121–128.

Salo, S, Storgårds, E and Wirtanen, G (1999) Alternative methods for sampling from surfaces. In: 30th R³-Nordic Contamination Control Symposium, VTT Symposium 193 (Eds. Wirtanen, G, Salo, S & Mikkola, A), Libella Painopalvelu Oy, Espoo. ISBN 951-38-5268-7, pp. 187–198.

Santegoeds, CM, Ferdelman, TG, Muyzer, G and de Beer, D (1998) Structural and functional dynamics of sulfate-reducing population in bacterial biofilms. *Applied and Environmental Microbiology* **64**:3731–3739.

Schadow, KH, Simpson, WA and Christensen, GD (1988) Characteristics of adherence to plastic tissue culture plates of coagulase—negative staphylococci exposed to subinhibitory concentrations of antimicrobial agents. *Journal of Infectious Diseases* **157**:71–77.

Schaule, G, Flemming, HC and Ridgway, HF (1993) Use of 5-cyano-2,3-ditolyl tetrazolium chloride for quantifying planktonic and sessile respiring bacteria in drinking water. *Applied and Environmental Microbiology* **59**:3850–3857.

Scheu, PM, Berghof, K and Stahl, U (1998) Detection of pathogenic and spoilage microorganisms in food with the polymerase chain reaction. *Food Microbiology* **15**:13–31.

Schmitt, J and Flemming, H-C (1996) FTIR spectroscopy. In: Microbial deterioration of materials (Eds. Sand, W, Heitz, E and Flemming, H-C), Springer Verlag, Heidelberg, pp. 143–157.

Sharma, DK and Prasad, DN (1992) Rapid identification of viable bacterial spores using a fluorescence method. *Biotechnology and Histochemistry* **67**:27–29.

Shotton, DM (1989) Confocal scanning optical microscopy and its applications for biological specimens. *Journal of Cell Science* **94**:175–206.

Sigsgaard, P (1989) Fluorescence microscopy techniques in the practice of food microbiology. In: Fluorescent Analysis in Foods (Ed. Munk, L), Longman, London, pp. 131–141.

Singh, A, Pyle, BH and McFeters, GA (1989) Rapid enumeration of viable bacteria by image analysis. *Journal of Microbiological Methods* **10**:91–101.

Stewart, PS, Griebe, T, Srinivasan, R, Chen, C-I, Yu, FP, de Beer, D and McFeters, GA (1994) Comparison of respiratory activity and culturability during monochloramine disinfection of binary population biofilms. *Applied and Environmental Microbiology* **60**:1690–1692.

Storgårds, E, Juvonen, R, Vanne, L and Haikara, A (1997) Detection methods in process and hygiene control. In: European Brewery Convention Monograph 26, Verlag Hans Carl, Getränke-Fachverlag, Nürnberg, pp. 95–107.

Storgårds, E, Salo, S, Yli-Juuti, P, Wirtanen, G and Haikara, A (1999a) Modern methods in process hygiene control—benefits and limitations. In: European Brewery Convention Congress 1999, Verlag Hans Carl, Getränke-Fachverlag, Nürnberg, pp. 249–258.

Storgårds, E, Simola, H, Sjöberg, A-M and Wirtanen, G (1999b) Hygiene of gasket materials used in food processing equipment. Part 1: New materials. The Transactions of the Institution of Chemical Engineers: Part C—Food and Bioproducts Processing, 77, pp. C2:137–145.

Storgårds, E, Simola, H, Sjöberg, A-M and Wirtanen, G (1999c) Hygiene of gasket materials used in food processing equipment: Part 2: Aged materials. The Transactions of the Institution of Chemical Engineers: Part C—Food and Bioproducts Processing, 77, pp. C2:146–155.

Tamachkiarowa, A and Flemming, HC (1996) Glass fibre sensor for the monitoring of biofouling. DECHEMA Monography 133:31–36.

Taylor, DL and Salmon, ED (1989) Basic fluorescence microscopy. In: Fluorescence Microscopy of Living Cells in Culture, Part A: Fluorescent Analogs, Labeling Cells and Basic Microscopy, Methods in Cell Biology, Vol. 29 (Eds. Wang, Y-L and Taylor, DL), Academic Press Inc., San Diego, pp. 207-237.

Thom, SM, Horobin, RW, Seidler, E and Barer, MR (1993) Factors affecting the selection and use of tetrazolium salts as cytochemical indicators of microbial viability and activity. Journal of Applied Bacteriology 74:433–443.

Thomas, JC, Desrosiers, M, St-Pierre, Y, Lirette, P, Bisaillon, JG, Beaudet, R and Villemur, R (1997) Quantitative flow cytometric detection of specific micro-organisms in soil samples using rRNA targeted fluorescent probes and ethidium bromide. Cytometry 27:224–232.

Tolker-Nielsen, T, Holmstrom, K, Boe, L and Molin, S (1998) Non-genetic population heterogeneity studies by in situ polymerase chain reaction. Molecular Microbiology 27:1099–1105.

Tolker-Nielsen, T, Holmstrom, K and Molin, S (1997) Visualization of specific gene expression in individual Salmonella typhimurium cells by in situ PCR. Applied and Environmental Microbiology 63:4196–4203.

Tuompo, H, Salo, S, Scheinin, L, Mattila-Sandholm, T and Wirtanen, G (1999) Chelating agents and detergents in sampling biofilm. In: Biofilms: The Good, the Bad and the Ugly (Eds. Gilbert, P, Brading, M, Walker, J and Wimpenny, J), BBC4, Bioline, Cardiff, ISBN 0-9520432-6-2, pp. 113–120.

Tuompo, H, Wirtanen, G, Salo, S, Scheinin, L, Båtsman, A and Levo, S (1998) Method and test kit for pretreatment of object surfaces, WO 98/07883, 18 August 1997, 26 February 1998, 33 p.

Ueckert, J, Breeuwer, P, Abee, T, Stephens, P, Nebe von Caron, G and ter Steeg, PF (1995) Flow cytometry applications in physiological study and detection of foodborne microorganisms. International Journal of Food Microbiology 28:317–326.

Wagner, M, Amann, R, Lemmer, H and Schleifer, KH (1993) Probing activated sludge with oligonucleotides specific for proteobacteria: inadecacy of culture-dependent methods for describing microbial community structure. Applied and Environmental Microbiology 59:1520–1525.

Walker, AJ, Holah, JT, Denyer, SP and Stewart, GSAB (1993) The use of bioluminescence to study the behaviour of Listeria monocytogenes when attached to surfaces. Colloids and Surfaces A: Physicochemical and Engineering Aspects 77:225–229.

Wallner, G, Erhart, R and Amann, R (1995) Flow cytomeric analysis of activated sludge with rRNA-targeted probes. Applied and Environmental Microbiology 61:1859–1866.

Wang, R-F, Cao, W-W and Cerniglia, CE (1997) A universal protocol for PCR detection of 13 species of foodborne pathogens in foods. *Journal of Applied Microbiology* **83**:727–736.

Wirtanen, G (1995) Biofilm formation and its elimination from food processing equipment, VTT Publication 251, VTT Offsetpaino, Espoo, Finland, 106 p. + app.48 p.

Wirtanen, G and Mattila-Sandholm, T (1993) Epifluorescence image analysis and cultivation of foodborne biofilm bacteria grown on stainless steel surfaces. *Journal of Food Protection* **56**:678–683.

Wirtanen, G and Mattila-Sandholm, T (1994) Measurement of biofilm of *Pediococcus pentosaceus* and *Pseudomonas fragi* on stainless steel surfaces. *Colloids and Surfaces B: Biointerfaces* **2**:33–39.

Wirtanen, G, Saarela, M and Mattila-Sandholm, T (2000) Biofilms—impact on hygiene in food industries. In: Biofilms II: Process Analysis and Applications (Ed. Bryers, J), John Wiley, New York, pp. 327–372.

Wirtanen, G, Salo, S, Maukonen, J, Bredholt, S and Mattila-Sandholm, T (1997) NordFood Sanitation in Dairies, VTT Publication 309, Espoo, VTT Offsetpaino, 47 p. + app 22 p.

Wirtanen, G, Nissinen, V, Tikkanen, L and Mattila-Sandholm, T (1995) Use of *Photobacterium leiognathi* in studies of process equipment cleanability. *International Journal of Food Science and Technology* **30**:523–533.

Yu, FP, Callis, GM, Stewart, PS, Griebe, T and McFeters, GA (1994) Cryosectioning of biofilms for microscopic examination. *Biofouling* **8**:85–91.

Yu, FP and McFeters, GA (1994a) Physiological responses of bacteria in biofilms to disinfection. *Applied and Environmental Microbiology* **60**:2462–2466.

Yu, FP and McFeters, GA (1994b) Rapid in situ assessment of physiological activities in bacterial biofilms using fluorescent probes. *Journal of Microbiological Methods* **20**:1–10.

Yu, FP, Pyle, BH and McFeters, GA (1993) A direct viable count method for the enumeration of attached bacteria and assessment of biofilm disinfection. *Journal of Microbiological Methods* **17**:167–180.

Zips, A, Schaule, G and Flemming, H-C (1990) Ultrasound as a means of detaching biofilms. *Biofouling* **2**:323–333.

4.3 Control of Biofilms in the Food and Beverage Industry

JOSEPH F. FRANK
Center for Food Safety and Quality Enhancement,
University of Georgia, Athens, GA, USA

INTRODUCTION

Biofilm control in the food and beverage industry is often required for both product contact and environmental surfaces. Although control of biofilms on product contact surfaces is critical to product quality, biofilms in the environment can become a source of pathogenic and spoilage microorganisms, being transferred to product or product contact surfaces by aerosol, implements and human contact. The degree of control required over environmental biofilms will depend on the degree to which the product is exposed to the environment after processing and the risk associated with contamination by spoilage or pathogenic microorganisms. For example, certain cheeses have a high degree of exposure to the environment during the ripening process and also, allow the growth of *Listeria monocytogenes*. Whereas, juices pasteurized in the retail container have no post-process exposure to the environment and because of their low pH they are at low risk for transmitting pathogenic bacteria.

Biofilm accumulation is the net result of attachment, growth at the surface and detachment. Control is achieved by limiting attachment and growth, while periodically promoting detachment through cleaning. Effective control in the food and beverage industry is usually achieved by using a combination of strategies. These include designing a processing environment that limits accumulation of product residues and microorganisms, maintaining nutrient, water and temperature conditions that limit biofilm development, periodic cleaning and sanitizing of product contact and environmental surfaces, and finally, monitoring the effectiveness of the control process. All these strategies should be employed as the limits of the process allow and as is necessary to produce a safe product with acceptable shelf life. For example, controlling water, nutrient accumulation and temperature on equipment surfaces during poultry processing is not

Industrial Biofouling. Edited by J. Walker, S. Surman, J. Jass

practical, so biofilm control in this situation depends primarily on equipment design and employing effective cleaning/sanitizing processes.

CONTROLLING MICROBIAL ATTACHMENT

Effective control of microbial attachment on food and beverage processing equipment is difficult. No process for inhibiting microbial attachment has demonstrated practical value in food processing. Attachment of specific microorganisms to food contact surfaces can be reduced by changing the physico-chemical properties of the surface. Adsorbed proteins can either increase or decrease attachment (Fletcher 1976; Helke *et al.* 1993; Carballo *et al.* 1991; Meadows 1971), presumably by their ability to change surface hydrophobicity and electrostatic charge (Al-Makhlafi *et al.*1994, 1995; Fletcher, 1976). High electrolyte concentrations in the surrounding medium can neutralize electrostatic forces involved in adhesion (Gordon and Miller 1998). Polymeric coatings can also provide nonspecific inhibition of microbial attachment (Humphries *et al.* 1987). Daeschael *et al.* (1992) and Bower *et al.* (1995) used antimicrobial proteins adsorbed onto a surface to inhibit attachment. Wood *et al.* (1996) generated a biocide at the surface to inhibit attachment. In some cases microbial attachment can also be reduced by increasing surface smoothness but this may not always be significant (Eginton *et al.* 1995).

The ability of the food and beverage industry to utilize attachment-resistant surfaces is limited by the requirement for durable non-reactive non-toxic materials. Stainless steel is not particularly resistant to microbial attachment but is the material of choice for most applications because of these other considerations. Microorganisms can rapidly attach to stainless steel surfaces and once established within the polishing grooves, are effectively removed only by using appropriate cleaning procedures (Zoltai *et al.* 1981; Stone and Zottolla 1985). Controlling attachment by modifying the surface is of also of limited value in the food industry because the material itself is not the attachment site, rather, attachment occurs on an organic or organic/mineral film that coats the surface soon after exposure to the product.

Equipment Design

Equipment design can influence microbial attachment. Attachment decreases with increasing shear forces. Therefore, designs that utilize continuous product flow, that reduce static product and have a minimum amount of valves, joints and other devices that provide protected sites, can result in reduced microbial attachment in the system. Since biofilms develop from microorganisms that adhere to a nutrient film, equipment design considerations for limiting accumulation of nutrient films are the same considerations that will limit microbial attachment.

Generally, food and beverage processing systems have not been designed or fabricated with a specific objective of limiting microbial attachment. Yet many of these systems operate without significant hygienic problems. This is because biofilm development is effectively controlled by other means. For inhibition of attachment to be an effective biofilm control measure, attachment must be greatly reduced or the effects of inhibition will be quickly overcome by subsequent growth. For example a 10-fold reduction in attachment will reduce microbial levels by only slightly more than three generation times. Just one attached microorganism, if given sufficient time, temperature and moisture conditions will become a microcolony that continuously sheds its progeny into the product or the processing environment. Observations that surface films of specific nutrients reduce attachment (Suarez et al., 1992; Carballo et al., 1991; Helke et al, 1993) do not imply that surfaces coated with such films will not readily support biofilm formation once the inevitable attachment occurs.

CONTROLLING GROWTH OF ATTACHED MICROORGANISMS

Biofilms will not develop unless conditions allow for microbial growth. Microbial growth requires water, nutrients, suitable temperature and sufficient time. Most food and beverage processing facilities contain specific growth niches in which all these conditions are present. Since few of these environments are sterile, biofilms will proceed to form. The underlying surface can affect the rate at which biofilms form (Rogers et al. 1994), however, materials used in the food industry will generally support extensive biofilm development in the presence of complex nutrients (Blackman and Frank 1996).

Limiting Water

Gabis and Faust (1988) discussed measures which can be taken to limit microbial growth in food processing environments. They conclude that controlling surface moisture is one of the most effective control measures. Guidelines for controlling the environment of environmental microorganisms in dairy plants also emphasize moisture control (USFDA-MIF 1988). Surfaces can be maintained as dry as possible by avoiding indiscriminant use of water, especially during clean-up, and by avoiding accumulation of condensation. Condensation on walls and equipment exteriors can be controlled by appropriate design and air dehumidification. Clean-up procedures should emphasize low pressure application of solutions and rinses. Exterior surfaces of equipment should be designed and equipment should be installed, so that water (or product) is not entrapped during processing or cleaning. Of special concern is the entrapment of water or product at inaccessible or hidden sites, since these areas are often not effectively cleaned. Examples of growth sites given by Gabis and Faust (1988) include: improperly welded metal surfaces with gaps, metal surfaces bolted together, improperly installed or drilled stainless steel cladding, conveyor parts, electrical boxes and equipment mounting brackets. After equipment clean-up, product contact surfaces should completely drain. If this is not possible, use of an acidified water rinse can help inhibit biofilm growth.

Food plants involved in handling dry products or utilizing dry processing methods (i.e. bakery, confectionery, powder blending) are usually not designed for wet cleaning. Water control is extremely important in these environments, since high amounts of nutrients are available as product dust. Sporadic water availability on a surface will result in microbial growth which is then spread throughout the plant by aerosolized dust when the surface dries. Facilities using both wet and dry processes must maintain strict physical separation between wet and dry areas, so that water is not transferred to the dry area, and nutrient-containing dust is not transferred to the wet area. If normally dry areas become wet, either through accident or because of cleaning activity, the site must be rapidly and thoroughly dried before microbial growth is initiated.

Limiting Nutrients

Reduction in available nutrients provides another means to control biofilm development. Product residues in food and beverage plants generally have sufficient nutrient content to support extensive microbial growth. Even condensation on clean walls and ceilings can support biofilm formation by absorption of nutrients from the air. Implementation of effective cleaning procedures can reduce nutrient availability and assist in biofilm control, both on environmental (non-food contact) surfaces and product contact surfaces. Although processing equipment is usually designed for efficient removal of food residues, environment surfaces and equipment exteriors often are not, and when moisture is supplied, they can harbor biofilms.

Removing nutrients will not result in as effective control over biofilm control as eliminating water. Low nutrient levels may enhance attachment of microorganisms. *Listeria monocytogenes* grown in a minimal medium (low in organic matter) showed 50-fold greater attachment to stainless steel than when grown in a complex medium (Kim and Frank 1994). Once attached, microorganisms are efficient at removing low levels of nutrients from the surrounding liquid. *L. monocytogenes* will readily produce biofilms at low levels of glucose, phosphate or nitrogen. Growth limiting levels of amino acids will delay initial biofilm growth of *L. monocytogenes* but after 12 days, these biofilms are just as extensive as those growing in more nutritious media (Kim and Frank 1995). Reducing nutrient levels at a surface may effectively delay biofilm formation to the extent that it can be controlled by periodic cleaning and sanitizing. Moist surfaces that are not periodically and effectively cleaned will ultimately develop a biofilm, even if there is no apparent accumulation of nutrients.

Effect of Temperature

Biofilms can form at most non-freezing temperatures found in food and beverage plants, with more rapid formation occurring as the temperature increases. Refrigeration temperatures can delay, but not stop, the formation of *L. monocytogenes* biofilms (Jeong and Frank 1994). If hot product continuously flows over a surface for periods greater than 16 hours, *Bacillus* spp. and other thermoduric microorganisms may form a biofilm (Hull *et al.* 1992; Lehman 1992). Langeveld *et al.* (1995) isolated various bacteria in biofilms on the interior wall of a heat exchanger that was heating milk to 80° C within 1.5 minutes. These included a coliform, a streptococcus and an aerobic spore former. The temperature of potential biofilm-forming sites is generally determined by the nature of the process, rather than by the need to control microbial growth. If adequate water and nutrient are present, surface temperature will be the major influence determining the required frequency of cleaning to achieve biofilm control. Surfaces that accumulate food residues may require cleaning at 2–4 hour intervals (ICMSF1980). Hot or warm product-contact surfaces may require cleaning every 14–18 hours, whereas, milk-contact surfaces at ambient temperature require cleaning after 16–20 hours of production. Instituting longer production runs to gain increased plant efficiency has often resulted in decreased product quality that can be traced back to biofilm accumulation.

REMOVAL OF BIOFILM AND PRODUCT RESIDUES

Microbial attachment and subsequent growth are inevitable in most food processing facilities. Therefore, to achieve sufficient biofilm control, effective removal of microorganisms and product residues through periodic cleaning is required. If biofilms are allowed to form, surfaces become more difficult to clean and subsequent sanitation efforts will be less effective. Therefore, cleaning processes that are both effective and sufficiently frequent are the most important aspect of biofilm control in the food and beverage industry.

Plant and Equipment Design

Many biofilm problems can be avoided through appropriate processing plant design. The principles of sanitary plant and equipment design have been reviewed by Hayes (1992) and Troller (1983). Of primary importance is a design that facilitates cleaning of all interior surfaces that can potentially harbour a biofilm. This means that walls, ceilings and floors must be constructed using smooth, non-porous material free of cracks and crevices. Tiled surfaces are suitable for floors and walls, as the grouted joints do not compromise their cleanability (Taylor and Holah 1996). Floors must completely drain, while floor drains should be designed for easy cleaning, since all provide excellent opportunities for biofilm growth. Metal grates covering floor drains are a common source of pathogen-containing biofilms. Condensation accumulation, a common source of biofilm problems, can often be controlled by employing designs that avoid exposed cold surfaces and adequate dehumidification of the air.

Faults in equipment design that can lead to biofilm development include lack of self-draining, entrapment of product and lack of accessibility for cleaning. Materials used to fabricate equipment must have durable, abrasion-resistant surfaces, as abrasions protect microorganisms from the cleaning process (Holah and Thorpe 1990; Stevens and Holah 1993) and also prevent complete removal of food residues (Masurovsky and Jordan 1958). Mechanically polished stainless steel generally meets requirements for equipment sanitation in regard to durability and is easier, or as easy, to clean as alternate materials such as aluminium, polymers, glass, mineral resins and porcelain (Boulange-Petermann 1996). However, faults in the polish surface have an adverse impact on cleanability because residues in pits and crevices are protected from cleaning shear forces (Leclercq-Perlat and Lalande 1994; Leclercq-Perlat et al. 1994; Freeman et al. 1990). Extra attention should be given to the design of product conveyor systems, especially belt conveyors, as many are not designed to be effectively cleaned or easily maintained. Pumps, valves and joints also commonly provide sites for biofilm formation because they are not designed for easy disassembly.

Well-designed equipment that is not properly maintained will rapidly fall into a state of uncleanability. Karpinsky and Bradley (1988) found that air-actuated butterfly valves required maintenance every 3–6 months to maintain cleanability. Cleanability failures were due to damage to the valve seats. Surface cracks in soft parts protect soil and microorganisms from the cleaning process (Major 1962). Microorganisms will attach to rubber and Teflon gaskets in dairy processing lines even when lines are properly cleaned (Austin and Bergeron 1995). In another study rubber accumulated ten times more milk-based soil than stainless steel (Dunsmore et al. 1981). Older gaskets allow progressively greater attachment and soil retention, so a periodic replacement programme is required. Stainless steel that has become pitted due to age or corrosion can be impossible to clean adequately using standard practices (Maxcy 1969).

Cleaning Processes

The objective of the cleaning process in the food and beverage industry, is to remove product residues and biofilms from equipment surfaces. Product-contact surfaces must achieve a greater degree of cleanliness than environmental surfaces. The basic components of a wet cleaning process include: (1) removal of gross soil using warm or cold water; (2) application of cleaning agent; (3) hand or mechanical scrubbing using a brush or turbulent flow; (4) rinsing off suspended soil and cleaning solution; (5) application of disinfectant; and (6) rinsing disinfectant residues from the surface (ICMSF 1980). Conventional cleaning processes used in the food and beverage industry were designed to remove product residues with no consideration for removal of biofilms. These processes may be adequate for many systems, especially if applied often enough to prevent biofilm accumulation. However, for some food processing systems, biofilm accumulation is inevitable and must be considered a target of the cleaning effort. Many cleaning protocols are designed with an unstated assumption that if food residues are sufficiently removed, biofilm will also be removed. Since biofilm is chemically distinct from food and beverage residues, it may be more difficult to remove, making this is a false assumption. Little research has been done in this area, however, evidence from Stone and Zottola (1985), Holah and Thorpe (1990) and Wirtanen et al. (1996) indicate that typical cleaning procedures do not remove all biofilm microorganisms or associated glycocalyx. Residual glycocalyx on a surface may increase subsequent attachment of microorganisms (Neu 1992) leading to more rapid biofilm accumulation. If a cleaning procedure does not thoroughly remove accumulated soils (biofilm or food), then repeated cleaning/ processing cycles will result in increasing soil accumulation with an eventual decrease in product quality or safety (Dunsmore et al. 1981).

Cleaning systems rely on chemical concentration, physical action, temperature and time to achieve their purpose. It is the interaction of these factors that will ultimately control the level of residual micro-organisms (Dunsmore 1981). Commonly used alkali cleaning agents will remove biofilm if applied in sufficient concentration, time and temperature (Wirtanen *et al.* 1996; Dunsmore 1981; Stone and Zottola 1985). Unfortunately, these conditions have not been defined for specific commercial processing systems. Arizcun *et al.* (1998) tested numerous treatments for removal of *L. monocytogenes* biofilms, with good results using sequential treatments with NaOH (pH 10.5) followed by acetic acid (pH 5.4), each applied at 55° C for 5 minutes. However, the model system used in this study was a pure culture biofilm formed in broth, perhaps easier to remove than the multispecies biofilms containing food residues or mineral deposits that exist in the industry. Evidence indicates that longer exposure times or greater detergent concentrations than normally used, may be required to remove all biofilm residual (Wirtanen *et al.*, 1996). Chlorinated alkali may be more effective at biofilm removal than unchlorinated and acidic cleaning agents are relatively ineffective (Czechowski 1991). Ethylene diaminetetraacetic acid (EDTA) is often added to cleaning solutions made from hard water. Wirtanen *et al.* (1996) found that the addition of EDTA enhanced biofilm removal. It is unclear if the benefit was due to chelation of water minerals, chelation of cations required for glycocalyx stability or a combination of these factors.

Cleaning efficacy is affected by water hardness, soil type and equipment design. Because of this complexity, cleaning systems should be designed by specialists to meet the needs of each processing facility while minimizing chemical and energy costs. For example, removal of animal fats requires cleaning solutions heated to 70° C or greater, whereas most non-fatty, uncooked-on soils can be removed using cleaning solutions at 43–54° C (Troller 1983). Prerinsing with hot water (>65° C) may cause denaturation of protein residues making them more difficult to remove. Control of cleaning solution temperature is critical, since excessive temperatures are uneconomical, whereas, inadequate heat may result in lack of soil removal or redeposition of dissolved soil.

Various cleaning methods are used in the food and beverage industry. These are: clean-in-place systems, washing machines, manual cleaning using foam, brushes or other scrubbing aids, and dry cleaning (ICMSF 1980). These methods have been adequately described by Hayes (1992) and Troller (1983). Although there is no data directly comparing of the relative effectiveness of these methods relative to biofilm removal, methods that rely less on manual scrubbing can be better controlled to give consistent results. Cleaning-in-place (CIP) systems are probably most effective for biofilm removal because they are automated and can employ stronger cleaning solutions and higher temperatures than manual systems. Unfortunately, poorly designed CIP systems can allow pockets of microbial growth which can go undetected because of lack of manual disassembly. Washing machines, on the other hand, require disassembly of the equipment that is then placed in the machine which circulates cleaning solution, with or without application of ultrasonic energy. After cleaning, the pieces are rinsed, sanitized and the equipment reassembled. This method can be very effective for biofilm removal if the washing machines are well maintained.

Equipment exteriors and environmental surfaces (walls, floors, etc.) are commonly cleaned by application of foam or gel. This method has replaced the use of high pressure sprays, which were found to spread microorganisms throughout the plant. The cleaning foam or gel clings to the target surface providing a long contact time. It can then be rinsed off using a low pressure spray or, if greater control over water is required, a vacuum system. Although there is a lack of research data on the effectiveness of foam/gel application for biofilm removal, it is an improvement over alternative methods (hand scrubbing, high pressure application) because irregular surfaces can be treated with efficient use of labor and with minimal aerosolization of biofilm microflora.

Manual cleaning is generally the least desirable wet cleaning method. Effectiveness depends on the reliability of the worker and the closeness of worker supervision. Milder cleaning solution concentrations and lower cleaning temperatures must be used. Cleaning implements can serve as sources of cross-contamination, especially if biofilms are present, as the biofilm microflora may not be inactivated by the cleaning solution. Complete removal of biofilm by manual cleaning is probably difficult (controlled observations have not been reported), but there are circumstances where manual scrubbing is required, especially when gross amounts of biofilm are present (Exner *et al.* 1987).

As previously discussed, water control is essential for biofilm control in normally dry process areas. In these cases, dry cleaning methods are superior to wet methods. It may be advisable for some dry processing facilities, such as chocolate manufacture or powder blending operations to never employ wet cleaning methods, as long as pathogen control is maintained. Dry cleaning utilizes vacuum collection of food residues, with scrapping of adherent residues when necessary. Care must be taken not to distribute dust throughout the plant while dry cleaning. If pathogens are found in the dust of a dry operation, the operation must be shut down for thorough wet cleaning, disinfection and redrying of the total environment. Such a process is time consuming, expensive and tedious while not always successful in eliminating the pathogen.

Cleaning Frequency

As previously discussed, the required frequency of cleaning to prevent biofilm growth in a wet environment will depend, to a great extent, on the surface temperature. Other factors include nutrient availability at the surface and thoroughness of the cleaning procedure, when it is applied. In some segments of the food and beverage industry, cleaning frequency is determined by the longest time a product can be processed before shelf life deteriorates. Unfortunately, this criterion does not consider public health consequences. For example, at the height of processing season, some frozen vegetable processors clean equipment every 10 days, with no noticeable effect on product shelf life. However, during this time *Listeria*-containing biofilms may develop on the equipment resulting in a large amount of product being exposed to low levels of this pathogen (unpublished data).

SURFACE DISINFECTION

Dunsmore *et al.* (1981) described the food contamination process in the following sequence: (1) deposition of microorganisms onto the equipment surface; (2) attachment of microorganisms to the surface; (3) removal of a portion of the attached population by the cleaning/sanitizing process; (4) growth of surviving microorganisms on the surface after cleaning; and (5) contamination of product when processing commences. This process may repeat with each use of the equipment. Both effective cleaning (soil/biofilm removal) and sanitizing (microbial inactivation), are required to break this cycle. The major challenge of the sanitizing process is to inactivate adherent microorganisms remaining after cleaning, whereas, the major challenge of the cleaning process is to leave as few of these microorganisms on the equipment surface as possible. Adherent microorganisms not removed by the cleaning/sanitizing process, are present in microscopic or macroscopic protected sites (Bai *et al.* 1968; Maxcy 1969), often referred to as 'harborages'. Microbial growth in these protected sites probably has the characteristics of a biofilm, including, if time allows, production of protective glycocalyx. Therefore, the ability of a sanitizing process to inactivate biofilm microorganisms is highly relevant to the microbial quality of many processed foods. Of course, with ideal equipment and plant design, there would be no harborages, the cleaning process would leave the equipment and processing environment perfectly clean and the sanitizing process would be unchallenged, having only to inactivate microorganisms exposed by the cleaning treatment. Practical experience indicates that food and beverage processing plants seldom approach this ideal situation. The realization that imperfect cleaning systems allow biofilm development in protected sites has led to research on the ability of chemical sanitizers used in the food industry to inactivate biofilm microorganisms and a search for improved sanitizing systems.

Effect of Underlying Surface

Surface materials used in the food industry differ in the ease with which they can be disinfected (Boulange-Petermann 1996). Polymeric (plastic) materials and aluminium require greater concentrations of sanitizer to inactivate adhering microorganisms than does stainless steel (Mafu et al. 1990; Kryzsinski et al., 1992; Gelinas and Goulet 1983). Stainless steel has also been found to be more easily disinfected than copper and carbon steel (Chen et al. 1993). Gasket materials (Teflon or rubber) are inherently difficult to sanitize (Mosteller and Bishop 1993; Ronner and Wong 1993). Surface abrasion can lead to less effective inactivation of adherent microorganisms on stainless steel and other materials. For example, electropolishing stainless steel provides a smooth surface that is easily disinfected but if this surface is abraded, it is significantly more difficult to disinfect (Frank and Chmielewki 1997).

Chemical Disinfectants

Chemical disinfectants commonly used in the food and beverage industry include chlorine (hypochlorous acid-producing compounds), quaternary ammonium compounds, anionic acid compounds and peroxyacetic acid. Less widely used sanitizing agents include iodophors, organic chlorine compounds, amphoteric surfactants and hydrogen peroxide. The general properties and use of these disinfectants have been reviewed by Cords and Dychdala (1993).

Various studies have investigated the ability of commonly used disinfectants to inactivate biofilms. The disinfectants most effective against suspended microorganisms are not necessarily most effective against the same microorganisms in a biofilm (Holah et al. 1990). Biofilms of L. monocytogenes treated with up to 800 quaternary ammonium compound and 400 ppm anionic acid disinfectant, were not completely inactivated after 20 minutes of exposure (Frank and Koffi 1990). Mostellar and Bishop (1993) found that all disinfectants studied, including iodophor, hypochlorite, acid anionic, peroxyacetic acid, fatty acid and quaternary ammonium compound, failed to inactivate bacteria attached to Teflon or rubber gaskets by more than 99.9% after 30 seconds of exposure. Chlorine, iodophor, phenolic and quaternary ammonium compounds failed to completely inactivate Pseudomonas biofilms on a polyvinyl chloride surface (Anderson et al. 1990). Wirtanen and Mattila-Sandholm (1992a) observed survival of various bacteria, including Pseudomonas fragi and L. monocytogenes in biofilms treated with hypochlorite, iodophor, quaternary ammonium compound and amphoteric disinfectant. The previously described studies used pure culture biofilms, which may not reflect reality, especially on equipment exteriors or when surfaces are exposed to raw product. Bourion and Cerf (1996) developed mixed culture biofilms of *Pseudomonas aeroginosa* and *L. innocua* and both microorganisms survived treatment with hypochlorite and peroxyacetic acid. Higher numbers of *L. innocua* survived disinfection in the mixed culture biofilm than in pure culture, indicating that a less prolific biofilm producer (*L. innocua*) can be protected by a more prolific biofilm producer.

Peroxide compounds may be more effective at inactivating biofilms than other disinfectants used in the food and beverage industry. Peroxyacetic acid is the most common of this type currently in use. Holah *et al.* (1990) found that peroxyacetic acid was the most effective of sanitizers tested for inactivating biofilms on stainless steel. Goroncy-Bermes and Gerresheim (1996) and Exner *et al.* (1987) also observed peroxyacetic acid to be effective against biofilms. However, research of Mostellar and Bishop (1993) indicated no increased efficacy for this sanitizer with regards to use on gasket biofilms. The effectiveness of peroxyacetic acid against biofilms may be due to its ability to maintain oxidizing potential in the presence of organic matter (Cords and Dychdala 1993). This is an especially useful property in the food industry, where biofilms will often contain large quantities of food residues in addition to microbial glycocalyx. Most of the model system biofilms used in the previously described studies probably contain less food residue than biofilms found in the industry. Therefore, sanitizers less adversely effected by organic matter may, in practice, be relatively more effective than the research indicates. Cords (1993) described a new peroxide disinfectant which is an equilibrium mixture of hydrogen peroxide, acetic acid and octanoic acid. Peroxidation of the octanoic acid forms a highly effective antimicrobial agent with greater hydrophobicity than other food industry disinfectants. This disinfectant could prove useful for penetrating food industry biofilms having a high lipid content due to the presence of lipophilic food residues or those formed by microorganisms that excrete hydrophobic polymers.

Disinfectants used in the food industry will, to various extents, leave behind attached glycocalyx material (Wirtanen and Mattila-Sandholm 1992a) and inactivated or injured cells (Mosteller and Bishop 1993; Schwach and Zottola 1984). As noted by Mattila-Sandholm and Wirtanen (1992), residual glycocalyx could enhance biofilm accumulation during subsequent soiling cycles, making effective cleaning more difficult.

Time of growth is an important factor determining the efficacy of biofilm disinfection treatments. *L. monocytogenes* allowed to attach for 4 hours can be completely inactivated by a quaternary ammonium compound and anionic acid disinfectant after 16 minutes, whereas, 14 day biofilms survived a 20 minute exposure (Frank and Koffi 1990). Eight day biofilms were over 100 times more resistant to 200 mg l^{-1} hypochlorite than were 4 hour attached cells (Lee and Frank 1991). Wirtanen and Mattila-Sandholm (1992b) observed that at least 48 hours was necessary for adherent microflora to develop sanitizer resistance. They associated this development with the formation of glycocalyx material. This helps explain why some investigators, studying inactivation of microorganisms attached to food contact surfaces, found that the cleaning/sanitizing treatments they employed were effective (Krysinski *et al.* 1992; Greene *et al.* 1993; Mustapha and Liewen 1989). These studies provided for 48 hours or less biofilm growth. The assumption behind these studies is that with effective cleaning every 24 hours, the survival characteristics of older biofilms are not relevant. This assumption does not hold up if harborages are present on the food contact surfaces of the processing equipment, with the consequences that biofilm is not completely removed with every cleaning, or if harbourages are present in the processing environment where cleaning may occur less often or be less effective. There is in agreement with that, like cleaning, the effectiveness of disinfection is reduced as the time interval between application increases.

Heat Disinfection

Heat is a widely used alternative to chemical disinfection. Heat is also used to sterilize processing systems. Biofilm microorganisms are more heat resistant than their planktonic counterparts. *L. monocytogenes* survived in biofilms heated to 70° C for 5 minutes (Frank and Koffi 1990). Heat in the form of hot water (>77° C) or steam is very effective at inactivating adherent microorganisms (ICMSF 1980), however, remaining food residues may be bonded to the underlying surface to such an extent that removal by subsequent cleaning will be very difficult. Therefore, disinfection by heat should not be used as a substitute for effective cleaning.

Fate of Detached Cells

Cells detached by the cleaning process are readily inactivated by chemical disinfectants. When clumps of *L. monocytogenes* biofilm were detached and dispersed, the cells were inactivated at a rate similar to their planktonic counterparts (Frank and Koffi 1990). Detached clumped cells, however, still maintained some resistance to the disinfectants. This data confirms the importance of the cleaning process for biofilm control.

DEVELOPMENT OF SANITIZER RESISTANT MICROFLORA

Biofilms develop in food and beverage processing environments in part because they impart sanitizer resistance to their constituent cells. Brown and Gilbert (1993) concluded that antimicrobial resistance of biofilms was not primarily due to glycocalyx because many antimicrobials readily diffuse through this barrier. Their conclusions were based primarily on research involving antibiotics. However, de Beer *et al.* (1994) determined that glycocalyx does impose a diffusion barrier on chlorine and that cell inactivation occurs along this barrier. This explains why detaching and breaking up clumps of biofilm cells increases their sensitivity to sanitizers and why survivors of a single cleaning/sanitizing process do not have increased resistance to these agents (Mattila *et al.* 1990). Although other mechanisms may be involved, current observations of model system biofilms indicate that resistance is due primarily to an inability of sanitizing agents to penetrate the glycocalyx.

A biofilm in a protected site within a food processing plant is subject to cycles of lethal/sublethal stress (from application of cleaning agents and disinfectants), followed by growth. Since the glycocalyx is a diffusion barrier, the surviving population will have been subject to an antimicrobial gradient. Such conditions are suitable for selection of a resistant population, especially since the sublethal exposure/growth cycle is repeated indefinitely. On the other hand, food industry disinfectants, such a chlorine and peroxides, have a non-specific mode of action which would be difficult to counteract through mutation. The scientific literature provides only hints of what might actually be occurring in a processing plant environment in this regard.

Heinzel (1988) distinguished three types of microbial resistance to disinfectants: (1) natural intrinsic resistance, ordinarily this resistance is conferred by biofilm growth as observed in laboratory model systems; (2) genetically acquired stable resistance; and (3) physiological adaptation, which is lost rapidly in the absence of the disinfectant. A fourth possibility not listed by Heinzel (1988), is the emergence of a resistant subpopulation. Pyle *et al.* (1994) isolated a halogen-resistant subpopulation of *Pseudomonas cepacia* from the space shuttle orbiter water system. The culture retained its resistance in the absence of selective pressure when it was grown on low nutrient media. Resistance was lost upon transfer to nutrient rich media. Halogen resistance is generally a result of aggregation, capsule production or other changes in the cell envelope. Since these mechanisms are under the physiological control of the cell, nutritional status of the cell can play a significant role in their expression. Bolton *et al.* (1988) isolated chlorine-resistant strains of *Staphylococcus aureus* from poultry processing plants. Resistance was associated with cell clumping, but could not be completely reversed by breaking up clumps using sonication. Non-clumping strains grown in media which enhanced polymer production gained some resistance but not to the level of the processing plant isolates. This study indicates that processing plant environments may select for microbial populations with stable resistance. Since aggregation is a trait of biofilm-forming cells, it is possible that the isolates developed their resistance as a result of repeated sanitizer exposure in a biofilm. Some isolates originated with the defeathering machinery, a site known to harbour large populations of microorganisms.

Resistance of *Pseudomonas aeruginosa* to quaternary ammonium compounds can be attributed to increased fatty acids and other lipophilic components in the cell wall (Sakagami *et al.* 1989). Although the authors did not determine if the resistance was under physiological control, it is apparent that to gain resistance, the cells must only selectively increase production of some existing cell wall components.

Resistance of *S. aureus* to quaternary ammonium compounds is plasmid-mediated. The resistance plasmid cloned by Tennent *et al.* (1985) was linked with antibiotic resistance, but the plasmid sequenced by Heir *et al.* (1995) was not. *S. aureus* strains harbouring the resistance plasmid were isolated from meat and poultry plants (Sundheim *et al.* 1992). Since such plasmids exist in food processing facilities, possibly in biofilm populations, there exists a potential for genetic transfer. Maillard *et al.* (1995) found that peracetic acid and sodium hypochlorite inhibited transduction of *P. aeruginosa*. Of course, food industry biofilms would not be constantly subject to transduction-inhibitory levels of disinfectants. Biofilms are the normal growth habitat for most microorganism, so genetic exchange within this habitat is expected. Recently, Christensen *et al.* (1998) observed conjugal transfer within a three species biofilm community. The implications of these observations for the food industry are yet to be determined.

MONITORING BIOFILM CONTROL

Biofilm control in the food and beverage industry cannot be successful without appropriate monitoring of control efforts. Because biofilms take time to develop and become more difficult to remove with time, early detection of conditions suitable for biofilm accumulation is required for successful control. Cleaning and sanitizer solution concentration, pH and temperature, should be closely monitored since deviations in these parameters can result in biofilm survival. Such monitoring is often a prerequisite program for facilities implementing Hazard Analysis Critical Control Point (HACCP) systems (NACMCF 1998).

Monitoring cleaned and sanitized equipment for biofilm survival is also necessary, because of the possible presence of protected sites or harborages. The critical aspect of monitoring for biofilm control is to sample sites that are at highest risk for biofilm development. It is from these protected sites that the biofilm will initiate growth when conditions are again suitable. Identifying proper sampling sites requires knowledge of equipment design, as well as, consideration of the conditions for microbial growth. Protected sites are often not readily accessible and are, therefore, inconvenient to sample. Microscopic-size protected sites might never be identified. Once high risk sites are identified, sampling can be accomplished with a sterile swab. Biofilm need not be quantitatively removed from the surface for monitoring to be effective. In fact, there is no practical means for quantitative sampling of biofilms on food processing equipment. Methods for detecting biofilms have been previously discussed. For monitoring purposes adenosine triphosphate analysis is preferred because it is rapid and determines both food and microbial residues. Plate count procedures are also commonly used. Monitoring equipment sanitation may be treated as either a critical control point or a prerequisite programme in the HACCP system, depending on the estimated risk to the public health of a sanitation failure.

REFERENCES

Al-Makhlafi, HA, McGuire, J and Daeschel, MA (1994) Influence of preadsorbed milk proteins on adhesion of *Listeria monocytogenes* to hydrophobic and hydrophilic surfaces. *Applied and Environmental Microbiology* **60**:3560–3565.

Al-Makhlafi, H, Nasir, A, McGuire, J and Daeschel, MA (1995) Adhesion of *Listeria monocytogenes* to silica surfaces after sequential and competitive adsorption of bovine serum albumin and -lactoglobulin. *Applied and Environmental Microbiology* **61**:2013–2015.

Anderson, RL, Holland, BW, Carr, JK, Bond, WW and Favero, MS (1990) Effect of disinfectants on pseudomonads colonized on the interior surface of PVC pipes. *American Journal of Public Health* **80**:17–21.

Arizcun, C, Vasseur, C and Labadie, JC (1998) Effect of several decontamination procedures on *Listeria monocytogenes* growing in biofilms. *Journal of Food Protection* **61**:731–734.

Austin, JW and Bergeron, G (1995) Development of bacterial biofilms in dairy processing lines. *Journal of Dairy Research* **62**:509–549.

Bai, B, Cousins, CM and Clegg, LF (1968) Studies on the laboratory soiling of milking equipment. *Journal of Dairy Research* **35**:247–256.

Blackman, IC and Frank, JF (1996) Growth of *Listeria monocytogenes* as a biofilm on various food-processing surfaces. *Journal of Food Protection* **59**:827–831.

Bolton, KJ, Dodd, CER, Mead, GC and Waites, WM (1988) Chlorine resistance of strains of *Staphylococcus aureus* isolated from poultry plants. *Letters in Applied Microbiology* **6**:31–34.

Boulange-Petermann, L (1996) Processes of bioadhesion on stainless steel surfaces and cleanability: A review with special reference to the food industry. *Biofouling* **10**:275–300.

Bourion, F and Cerf, O (1996) Disinfection efficacy against pure-culture and mixed-population biofilms of *Listeria innocua* and *Pseudomonas aeruginosa* on stainless steel, Teflon, and rubber. *Sciences des Aliments* **16**:151–166.

Bower, CK, McGuire, J and Daeschel, MA (1995) Suppression of *Listeria monocytogenes* colonization following adsorption of nisin onto silica surfaces. *Applied Environmental Microbiology* **61**:992–997.

Brown, MRW and Gilbert, P (1993) Sensitivity of biofilms to antimicrobial agents. *Journal of Applied Bacteriology Symposium Supplement* **74**:87S–97S.

Carballo, J, Ferreiros, CM and Criado, MT (1991) Importance of experimental design in the evaluation of proteins in bacteria adherence to polymers. *Medical Microbiology and Immunology* **180**:149–155.

Chen, CI, Griebe, T, Srinivasan, R and Stewart, P (1993) Effects of various metal substrata on accumulation of *Pseudomonas aeruginosa* biofilms and the efficacy of monochloramine as a biocide. *Biofouling* **7**:241–251.

Christensen, BB, Sternberg, C, Andersen, JB, Eberl, L, Moller, S, Givskov, M and Molin, S (1998) Establishment of new genetic traits in a microbial biofilm community. *Applied and Environmental Microbiology* **64**:2247–2255.

Cords, BR (1993) New peroxyacetic acid sanitizer. Proceedings of the Institute of Brewing 23rd Annual Meeting, pp. 165–169.

Cords, BR and Dychdala, GR (1993) Sanitizers: halogens, surface-active agents, and peroxides. In: Antimicrobials in Food (Eds. Davidson, PM and Branen, AL), Marcel Dekker, New York, pp. 469–537.

Czechowski, MH (1991) Biofilms and surface sanitation in the food industry. *Biodeterioration and Biodegradation* **8**:453–454.

Daeschel, MA, McGuire, J and Al.-Makhlafi, H (1992) Antimicrobial activity of nisin adsorbed to hydrophilic and hydrophobic silicon surfaces. *Journal of Food Protection* **55**:731–735.

de Beer, D, Srinivasan, R and Stewart, PS (1994) Direct measurement of chlorine penetration into biofilms during disinfection. *Applied and Environmental Microbiology* **60**:4339–4344.

Dunsmore, DG (1981) Bacteriological control of food equipment surfaces by cleaning systems. I. Detergent effects. *Journal of Food Protection* **44**:15–20.

Dunsmore, DG, Twomey, A, Whittlestone, WG, and Morgan, HW (1981) Design and performance of systems for cleaning product-contact surfaces of food equipment: A review. *Journal of Food Protection* **44**:220–240.

Eginton, PJ, Gibson, H, Holah, J, Handley, PS and Gilbert, P (1995) The influence of substratum properties on the attachment of bacterial cells. *Colloids and Surfaces B-Biointerfaces* **5**:153–159.

Exner, M, Tuschewitzki, GJ and Scharnagel, J (1987) Influence of biofilms by chemical disinfectants and mechanical cleaning. *Zentralblatt für Bakteriologie und Hygeine* **B183**:549–563.

Fletcher, M (1976) The effects of proteins on bacterial attachment to polystyrene. *Journal of General Microbiology* **94**:400–404.

Frank, JF and Chmielewski, RAN (1997) Effectiveness of sanitation using quaternary ammonium compound and chlorine on stainless steel and other domestic food preparation surfaces. *Journal of Food Protection* **60**:43–47.

Frank, JF and Koffi, RA (1990) Surface adherent growth of *Listeria monocytogenes* is associated with increased resistance to surfactant sanitizers and heat. *Journal of Food Protection* **53**:550–554.

Freeman, WB, Middis, J and Muller-Steinhagen, HM (1990) Influence of augmented surfaces and of surface finish on particulate fouling in double pipe heat exchangers. *Chemical Engineering Processing* **27**:1–11.

Gabis, D and Faust, RE (1988) Controlling microbial growth in food processing environments. *Food Technology* **42**:81–82, 89.

Gélinas, P and Goulet, J (1983) Efficacité de huit désinfectants sur trois types de surfaces contaminées par *Pseudomonas auruginosa*. *Canadian Journal of Microbiology* **29**:1716–1730.

Gordon, AS and Miller, FJ (1984) Electrolyte effects on attachment of an estuarine bacterium. *Applied and Environmental Microbiology* **47**:495–499.

Goroncy-Bermes, P and Gerresheim, S (1996) Efficacy of peroxide-containing solutions against microorganisms in biofilms. *Zentralblatt für Hygiene und Unweltmedizin* **198**:473–477.

Greene AK, Few BK, and Serafini JC (1993) A comparison of ozonation and chlorination for the disinfection of stainless steel surfaces. *Journal of Dairy Science* **76**:3617–3620.

Hayes, PR (1992) Food Microbiology and Hygiene, Elsevier Science Publishers Ltd., Essex, London, pp. 224–310.

Heinzel, M (1988) The phenomena of resistance to disinfectants and preservatives. In: Industrial Biocides (Ed. Payne KR), John Wiley and Sons, New York, pp. 52–67.

Heir, E, Sundheim, G and Holck, AL (1995) Resistance to quaternary ammonium compounds in *Staphylcoccus* spp. isolated from the food industry and nucleotide sequence of the resistance plasmid pST827. *Journal of Applied Bacteriology* **79**: 149–156.

Helke, DM, Somers, EB and Wong, ACL (1993) Attachment of *Listeria monocytogenes* and *Salmonella typhimurium* to stainless steel and buna-N in the presence of milk and individual milk components. *Journal of Food Protection* **56**:479–484.

Holah, JT, Higgs, C, Robinson, S, Worthington, D and Spenceley, H (1990) A conductance-based surface disinfection test for food hygiene. *Letters in Applied Microbiology* **11**:255–259.

Holah, JT and Thorpe, RH (1990) Cleanability in relation to bacterial retention on unused and abraded domestic sink materials. *Journal of Applied Bacteriology* **69**:599–608.

Hull, R, Toyne, S, Haynes, I and Lehman, F (1992) Thermoduric bacteria: A re-emerging problem in cheesemaking. *Australalian Journal of Dairy Technology* **47**:92–94.

Humphries, M, Nemcek, J, Cantwell, JB and Gerrard, JJ (1987) The use of graft copolymers to inhibit adhesion of bacteria to solid surfaces. *FEMS Microbial Ecology* **45**:297–304.

International Commission on Microbiology Specifications for Foods (1980) Microbiol Ecology of Foods. Vol. 1 Factors Affecting the Life and Death of Microorganisms. Academic Press, New York, pp. 232–258.

Jeong, DK and Frank, JF (1994) Growth of *Listeria monocytogenes* at 10° C in biofilms with microorganisms isolated from meat and dairy processing environments. *Journal of Food Protection* **57**:567–586.

Karpinsky, JL and Bradley, RL Jr. (1988) Assessment of the cleanability of air-actuated butterfly valves. *Journal of Food Protection* **51**:364–368.

Kim, KY and Frank, JF (1994) Effect of growth nutrients on attachment of *Listeria monocytogenes* to stainless steel. *Journal of Food Protection* **57**:720–726.

Kim, KY and Frank, JF (1995) Effect of nutrients on biofilm formation by *Listeria monocytogenes* on stainless steel. *Journal of Food Protection* **58**:24–28.

Krysinski, EP, Brown, LJ and Marchisello, TJ (1992) Effect of cleaners and sanitizers on *Listeria monocytogenes* attached to product contact surfaces. *Journal of Food Protection* **55**:246–251.

Langeveld, LP, Vanmontfortquasig, RM, Weerkamp, AH, Waalewijn, R and Wever, JS (1995) Adherance, growth and release of bacteria in a tube heat-exchanger for milk. *Netherlands Milk and Dairy Journal* **49**:207–220.

Leclercq-Perlat, MN and Lalande, M (1994) Cleanability in relation to surface chemical composition and surface finish of some materials commonly used in food processing. *Journal of Food Engineering* 23:501–517.

Leclercq-Perlat, MN, Tissier, JP and Benezech, T (1994) Cleanability of stainless steel in relation to chemical modifications due to industrial cleaning procedures used in the dairy industry. *Journal of Food Engineering* 23:449–465.

Lee, SH and Frank, JF (1991) Inactivation of surface-adherent *Listeria monocytogenes* by hypochlorite and heat. *Journal of Food Protection* 54:4–6.

Lehmann, FL (1992) Thermoduric-thermophilic bacteria in continuous cheese-making. *Australian Journal of Dairy Technology* 47:92–94.

Mafu, AA, Roy, D, Goulet, J, Savoie, L and Roy, R (1990) Efficacy of sanitizing agents for destroying *Listeria monocytogenes* on contaminating surfaces. *Journal of Dairy Science* 73:3428–3432.

Maillard, JY, Begg, TS, Day, MJ, Hudson, RA and Russell, AD (1995) Effects of biocides on the transduction of *Pseudomonas aeruginosa* PAO by bacteriophage F116. *Letters in Applied Microbiology* 21:215–218.

Major, WCT (1962) Sterilization of milking machine rubberware. *Queensland Journal of Agricultural Science* 19:117–121.

Masurovsky, EB and Jordan, WK (1958) Studies on the relative bacterial cleanability of milk contact surfaces. *Journal of Dairy Science* 41:1342–1358.

Mattila, T, Manninen, M and Kylasiurola, AL (1990) Effect of cleaning-in-place disinfectants on wild bacterial strains isolated from a milking line. *Journal of Dairy Research* 57:33–39.

Mattila-Sandholm, T and Wirtanen, G (1992) Biofilm formation in the industry: A review. *Food Reviews International* 8:573–603.

Maxcy, RB (1969) Residual microorganisms in cleaned-in-place systems for handling milk. *Journal of Milk and Food Technology* 32:140–143.

Meadows, PS (1971) The attachment of bacteria to solid surfaces. *Archives of Microbiology* 75:374–381.

Mosteller, TM and Bishop, JR (1993) Sanitizer efficacy against attached bacteria in a milk biofilm. *Journal of Food Protection* 56:34–41.

Mustapha, A and Liewen, MB (1989) Destruction of *Listeria monocytogenes* by sodium hypochlorite and quaternary ammonium sanitizers. *Journal of Food Protection* 52:306–311.

National Advisory Committee on Microbiological Criteria for Foods (1998) Hazard analysis and critical control point principles and application guidelines. *Journal of Food Protection* 63:1246–1259.

Neu, TR (1992) Microbial 'footprints' and the general ability of microorganisms to label interfaces. *Canadian Journal of Microbiology* 38:1005–1008.

Pyle, BH, Watters, SK and McFeters, GA (1994) Physiological aspects of disinfection resistance in *Pseudomonas cepacia*. *Journal of Applied Bacteriology* 76:142–148.

Rogers, J, Dowsett, AB, Dennis, PJ, Lee, JV and Keevil, CW (1994) Influence of temperature and plumbing material selection on biofilms formation and growth of *Legionella pneumophila* in a model potable water system containing complex microbial flora. *Applied and Environmental Microbiology* 60:1585–1592.

Ronner, AB and Wong, ACL (1993) Biofilm development and sanitizer inactivation of *Listeria monocytogenes* and *Salmonella typhimurium* on stainless steel and Buna-n rubber. *Journal of Food Protection* 56:750–758.

Sakagami, Y, Yokoyama, H, Nishimura, H, Ose, Y and Tashima, T (1989) Mechanism of resistance to benzalkonium chloride by *Pseudomonas aeruginosa*. *Applied and Environmental Microbiology* 55:2036–2040.

Schwach, TS and Zottola, EA (1984) Scanning electron microscopic study on some effects of sodium hypochlorite on attachment of bacteria to stainless steel. *Journal of Food Protection* **47**:756–759.

Stevens, RA and Holah, JT (1993) The effect of wiping and spray-wash temperature on bacterial retention on abraded domestic sink surfaces. *Journal of Applied Bacteriology* **75**:91–94.

Stone, LS and Zottola, EA (1985) Effect of cleaning and sanitizing on the attachment of *Pseudomonas fragi* to stainless steel. *Journal of Food Science* **50**:951–956.

Suarez, B, Ferreiros, CM and Criado, MT (1992) Adherance of psychrotrophic bacteria to dairy equipment surfaces. *Journal of Dairy Research* **59**:381–388.

Sundheim, G, Hagtvedt, T and Dainty, R (1992) Resistance of meat associated staphylcocci to a quaternary ammonium compound. *Food Microbiology* **9**:161–167.

Taylor, JH and Holah, JT (1996) A comparative evaluation with respect to the bacterial cleanability of a range of wall and floor surface materials used in the food industry. *Journal of Applied Bacteriology* **81**:262–266.

Tennent, JM, Lyon, BR, Gillispie, MF, May, JW and Skurray, RA (1985) Cloning and expression of *Staphylococcus aureus* plasmid-mediated quaternary ammonium compound resistance in *Escherichia coli*. *Antimicrobial Agents and Chemotherapy* **27**:79–85.

Troller, JA (1983) Sanitation in Food Processing, Academic Press, New York, pp. 21–110.

U.S. Food and Drug Administration, Milk Industry Foundation/International Ice Cream Association (1988) Recommended guidelines for controlling environmental contamination in dairy plants. *Dairy and Food Sanitation* **8**:52–56.

Wirtanen, G, Husmark, U and Mattila-Sandholm, T (1996) Microbial evaluation of the biotransfer potential from surfaces with *Bacillus* biofilms after rinsing and cleaning procedures in closed food-processing systems. *Journal of Food Protection* **59**:727–733.

Wirtanen, G and Mattila-Sandholm, T (1992a) Removal of foodborne biofilms— Comparison of surface and suspension tests Part I. *Lebsmittel Wissenschaft und Technologie* **25**:43–49.

Wirtanen, G and Mattila-Sandholm, T (1992b) Effect of the growth phase of foodborne biofilms on their resistance to chlorine sanitizer. Part II. *Lebsmittel Wissenschaft und Technologie* **25**:50–54.

Wood, P, Jones, M, Bhakoo, M and Gilbert, P (1996) A novel strategy for control of microbial biofilms through generation of biocide at the biofilm-surface interface. *Applied and Environmental Microbiology* **62**:2598–2602.

Zoltai, PT, Zottola, EA, and McKay, LL (1981) Scanning electron microscopy of microbial attachment to milk contact surfaces. *Journal of Food Protection* **44**:204–208.

5 Future Direction of Biofilm Research; Methodologies and their Applications

J. T. WALKER[1] AND J. JASS[2]
[1]*Microbiologist, Pathogen Microbiology Group, CAMR, Porton Down, Salisbury, SP4 0JG, UK*
[2]*Project leader, Department of Microbiology, Umeå University, SE-901 87 Umeå, Sweden*

BIOFILM RESEARCH—THE FUTURE DIRECTION

Research into microbial biofilms has increased over the past decade, as observed by the number of publications, books and conferences on this subject. Perhaps this is as much a reflection in terminology, as it is in the interest of biofilms *per se*, as not all surface associated bacteria have been thought of as biofilms. For example, in environments such as the mouth, 'plaque' is still the preferred word of use but it is in essence a biofilm.

STRUCTURE AND COMMUNICATION

Even though the volume of research into biofilms has increased over the past few years they are still poorly understood. In order to provide techniques and methods for biofilm measurement, scientists require an understanding of the fundamental structure and function of biofilms. We need to be able to understand more clearly what factors determine microbial biofilm growth and how this relates to its structure.

The question of whether microbial cells need to communicate with each other for growth has been a focus of much attention (Kaprelyants and Kell 1996). For several years it has been understood that tissue cells communicate with each other, but only recently have there been indications that bacteria also send and receive information among themselves. The best characterized example of interbacterial signalling is autoinduction (or quorum sensing) of the symbiotic marine bacterium *Vibrio fischeri* in the

Industrial Biofouling. Edited by J. Walker, S. Surman, J. Jass

light organs of certain marine fishes. The autoinduction of luminescence in
V. fischeri was described in the early 1970s by Nealson *et al.* (1970) and
Eberhard (1972). *V. fischeri* accumulates in the fish organ at high cell
densities (10^{10}–10^{11} cells per ml) and produces a small diffusible compound
called the autoinducer. At a critical concentration of the autoinducer, the *lux*
genes are activated producing the characteristic luminescence. Autoinduction only occurs where *V. fischeri* reaches high cell densities and these may
be found in biofilms where the cell to cell association is high but not in the
low population density of the planktonic state. In recent years a growing
number of Gram-negative bacteria such as *Pseudomonas aeruginosa* have
been discovered to have genes, similar to the *Lux* genes, that are regulated
by autoinducer molecules produced at high bacterial densities. These genes
are also associated with the regulation of a number of virulence factors
(Pearson *et al.* 1994).

Quorum sensing may not be limited to sending signals to cells of the
same species but may also send signals to other bacterial species. This may
help monitor and control the diversity of species or regulate intergeneric
communication in addition to communicating with in the species. It has
been suggested that quorum sensing may play an important role in the
development of biofilms since this environment is characterized by both
high cell densities and the close proximity of different species (Williams
and Stewart 1994).

Starting with single cells attaching to a surface, biofilms develop to
mature microcolonies with complex structures consisting of stacks and
aggregates of microbial cells (Costerton *et al.* 1995). This structure may have
a profound importance in the placement of particular species within the
biofilm, possibly dependent on nutrient conditions. The survival of
anaerobic species such as sulphate-reducing bacteria may indeed use
quorum sensing as a mechanism of cell–cell communication to sense
anaerobic sites within the biofilm. Whether these cells track their way
through the biofilm until they find a suitable anaerobic region or they are
within the aggregates that form part of a biofilm and only start to grow once
the region has become sufficiently reduced is unknown. In general,
communication signals produced by one bacteria might function as an
attractant for a second microorganism thus leading to the development of
mixed cultures functioning co-operatively within a particular ecological
niche. This is one way that quorum sensing may play a role in the control of
gene expression within biofilms. The ability of an organism to self sense
its own or other bacterial population densities may facilitate the initiation
of glycocalyx formation. Alternatively, it may lead to the induction of
other genes essential for the maintenance of biofilm growth, sloughing,
oxygenation or even reduction in the immediate locality (Williams and
Stewart 1994). Other bacteria may use such communication mechanisms to

aid transfer of nutrients and may influence the effectiveness of antimicrobial compounds.

Bacteria such as *P. aeruginosa* are a problem since they are known to form biofilms in both medical infections and industrial environments causing fouling and corrosion in water systems. These organisms have been shown to operate a quorum sensing system that is involved in the development of structurally complex biofilms (Stoodley *et al.* 1997). The involvement of an intercellular signal molecule in *P. aeruginosa* biofilm formation suggests a possible target for controlling biofilm growth (Davies *et al.* 1998). This has relevance in biofilms on catheters, in cystic fibrosis, and in industrial environments where *P. aeruginosa* biofilms are a persistent problem.

There is a large number of Gram-negative bacteria involved in biofilm formation and their growth may be affected by autoinduction and quorum sensing. It is clear that much work needs to be carried out in this area to understand the role that quorum sensing plays in biofilm growth and cell–cell, species–species interaction

WHEN IS A CELL VIABLE BUT NON-CULTURABLE?

Accurately determining the actual number of cells present is one of the anomalies of current microbiological practice (Barer 1997). One method used is plate culture where the number of colony forming units (cfu) indicates the number of cells able to grow using a particular agar media, however, it does not allow enumeration of 'dormant' cells. Gram-positive spore forming bacteria will, under unfavourable conditions, form spores that can become culturable cells at some condition-dependent point. Alternately, Gram-negative, non-spore forming organisms may reach a state known as viable but non-culturable (VNC) and these organisms cannot be recovered by routine isolation methods. Researchers have postulated that the VNC state may be the inter-epidemic reservoir of a number of pathogens including *Vibrio cholerae* (Roszak and Colwell 1987), *Enterococcus faecalis* (Lleo *et al.* 1998), *Escherichia coli* (Colwell *et al.* 1985), *Salmonella enteriditis* (Roszak *et al.* 1984), *Campylobacter jejuni* (Rollins and Colwell 1986), *Helicobacter pylori* (Kurokawa *et al.* 1999; Cellini *et al.* 1998) and *Legionella pneumophila* (Steinert *et al.* 1997; Hussong *et al.* 1987).

Quantitative assessment of bacterial contamination is usually carried out on the aqueous phase of mainly water samples. Little attention is given to the number of bacteria that may be attached to surfaces from which the water samples were taken, thus providing no information on the accumulated biofilms. Pathogenic bacteria may enter a state in which they retain their viability but cannot readily be cultured, inhibiting their isolation. Although the VNC state is still rather poorly understood, a

number of research programmes have advanced our knowledge in the area
(Barer *et al*. 1993; Colwell *et al*. 1985). Biofilms may present an environment
where the VNC mode of growth may provide a physiological advantage for
survival. In such a way, small numbers of microbial cells may evade
detection by standard culture techniques.

As cells become non-viable they may play an important role in nutrient
cycling or the production of secondary metabolites that inhibit competing
organisms (Barer and Harwood 1999). In another way, the death of cells
within a biomass may provide essential nutrients to support the growth of
other cells in the community enabling the fittest cells to survive. Further
research into whether biofilm bacteria may adapt a VNC mode of growth
may assist in developing control mechanisms of bacterial biofilms.

In industry and food and water environments, microbiologists have
identified that many problems of corrosion (Hamilton 1985; Little *et al*.
1991), oil souring (Bass *et al*. 1993), soiling and pipe/filter blocking
(Daschner *et al*. 1996) have been microbially derived. However, it has
often been difficult to persuade engineers and industrial management that
the underlying cause has been bacterial. A basic understanding of the
microbiology of biofilms must be available to a wider range of disciplines.
In the food industry, the use of Hazardous and Critical Control Points
(HACCP) has brought about a revolution in the use of microbiological
analysis for the control of hygiene in the work place and of the clean-in-
process (CIP) rinse systems. Here the benefits of understanding biofilm
formation, biofouling and their control have lead to longer food product
shelf life by controlling the presence of food pathogens and spoilage
microorganisms.

BIOFILM DETECTION

Traditional techniques used for monitoring biofilm formation include
the use of direct contact plates to enable the assessment of total bacterial
numbers and the identification of species to be enumerated. Contact plates,
however, have their limitations, including the long incubation periods for
the organisms to grow, need for qualified persons to interpret results and
the failure to detect VNC organisms which may be present. Therefore, there
is a necessity for improving methods and simplifying microbial and biofilm
detection within industry.

One of the techniques that has straddled the void between the laboratory
and the work place has been rapid adenosine triphosphate analysis.
Currently, a number of companies are producing kits that enable the on site
workforce to determine the extent of fouling of a surface. Such kits combine
ease of use, speed and accuracy with state-of-the-art data management and

trend analysis capabilities. Although this method determines total fouling rather than distinguishes between pathogenic and/or bacterial fouling, it does provide an immediate result of one's effort to reduce soiling on a surface. This technology has developed from large expensive bench-top mounted bioluminescent detectors to small affordable portable luminometers which have overcome the problems of reagent preparation and reading.

The technology still has limitations, including detection limits in the order of $\times 10^{3}$ cfu per cm^2, however these will improve as the technology advances. Rapid water test kits have been marketed to efficiently monitor microorganisms in traditional industrial systems such as cooling towers, recirculatory and closed water systems (Biotrace Limited). For biofilm assessment, the biological material still has to be removed from the substrata, requiring access to the fouling site and is still destructive to the biofilm structure.

Many other techniques used for determining microbial fouling also require access to the site where the fouling occurs to assess the microbiological loading of a surface. There are, however a number of side-stream devices that allow biofilms to be easily collected for analysis (see Chapter 3, section 2; Chapter 4, section 2). One of the most prolifically used engineering devices is the Robbins device. It has also been copied and modified for small scale pilot studies. These types of devices provide convenient mechanisms of presenting a substrata for biofilm development at the same time, using the environment created within the industrial system. However for analysis, this substrata has to be removed and the biofilm integrity disrupted.

Ideally, methods are required that allow on-line and in-line sensors that will monitor the presence and development of the biofilms. This technology will enable downstream data to be collected so that decisions can be made on the basis of real-time results of the developing biofilms. Thus process systems can be operated until a problem is identified by the in-line instrumentation, rather than sacrificing the integrity of pipe work to determine the presence and extent of the biofouling. Such equipment must be rapid, able to measure and differentiate between planktonic and biofilm cells, simple to operate, accurate, reliable and robust.

Although we have made some progress in understanding and detecting biofilms in industrial processes and environments, controlling biofilm formation has been very difficult. Thus research needs to continue to develop new methodology for the detection of microbial biofilms and fouling in different industries. For the future, advanced technologies are required to be developed that will allow in-line, on-line measurement of biofilms, relaying data down-stream where alarms can be activated when biofouling problems arise. However, currently there is still much research to be carried out to provide us with a better understanding of the

Table 5.1. Techniques from which in-line and on-line biofilm detectors may be developed

Technique	Application	Reference
Optical Monitoring	Beverage/water industry	(Flemming and Schaule 1996a, b; Tamachkiarowa and Flemming 1996)
Electrochemical	Environmental/aggregates	(Bento *et al.* 1998)
Heat transfer resistance	Industrial heat exchangers	(Melo and Pinheiro 1992)
Ion mobility techniques	Process water	(Paakkanen and Huttunen 1994; Kotiaho *et al.* 1995)
Infrared techniques FTIR ATR	Pharmaceutical/clean water technologies	(Nichols *et al.* 1985) (Nivens *et al.* 1993a, b, c)

FTIR, Fourier transform infrared spectroscopy; ATR, attenuated total reflection spectroscopy.

fundamental control parameters in the structure and communication mechanisms of bacterial cells within biofilms.

REFERENCES

Barer, MR (1997) Viable but non-culturable and dormant bacteria: time to resolve an oxymoron and a misnomer? *Journal of Medical Microbiology* **46**:629–631.

Barer, MR, Gribbon, LT, Harwood, CR and Nwoguh, CE (1993) The viable but non-culturable hypothesis and medical bacteriology. *Reviews in Medical Microbiology* **4**:183–191.

Barer, MR and Harwood, CR (1999) Bacterial viability and culturability. *Advances in Microbial Physiology* **41**:94–137.

Bass, CJ, Lappin-Scott, HM and Sanders, PF (1993) Bacteria that sour reservoirs: new concept for the mechanisms of reservoir souring by sulphide generating bacteria. *Journal of Offshore Technology* **1**:31–37.

Bento, MF, Thouin, L, Amatore, C and Montenegro, MI (1998) About potential measurements in steady state voltammetry at low electrolyte/analyte concentration ratios. *Journal of Electroanalytical Chemistry* **443**:137–148.

Cellini, L, Robuffo, I, Di Campli, E, Di Bartolomeo, S, Taraborelli, T and Dainelli, B (1998) Recovery of *Helicobacter pylori* ATCC43504 from a viable but not culturable state: regrowth or resuscitation? *Apmis* **106**:571–579.

Colwell, RR, Brayton, PR, Grimes, DJ, Roszak, DB, Huq, SA and Palmer, LM (1985) Viable but non-culturable *Vibrio cholerae* and related pathogens in the environment: implications for release of genetically engineered microorganisms. *Biotechnology* **3**:817–820.

Costerton, JW, Lewandowski, Z, Caldwell, DE, Korber, DR and Lappin-Scott, HM (1995) Microbial biofilms. *Annual Reviews in Microbiology* **49**:711–745.

Daschner, FD, Ruden, H, Simon, R and Clotten, J (1996) Microbiological contamination of drinking water in a commercial household water filter system. *European Journal of Clinical Microbiology and Infectious Diseases* **15**:233–237.

Davies, DG, Parsek, MR, Pearson, JP, Iglewski, BH, Costerton, JW and Greenberg, EP (1998) The involvement of cell-to-cell signals in the development of a bacterial biofilm. *Science* **280**:295–298.

Eberhard, A (1972) Inhibition and activation of bacterial luciferase synthesis. *Journal of Bacteriology* **109**:1101–1105.

Flemming, H-C and Schaule, G (1996a) Biofouling. In: Microbial deterioration of materials (Eds. Sand, W, Heitz, E and Fleming, H-C), Springer Verlag, Heidleberg, pp. 39–54.

Flemming, H-C and Schaule, G (1996b) Measures against biofouling. In: Microbial deterioration of materials (Eds. Sand, W, Heitz, E and Fleming, H-C), Springer Verlag, Heidleberg, pp. 121–139.

Hamilton, WA (1985) Sulphate-reducing bacteria and anaerobic corrosion. *Annual Reviews in Microbiology* **39**:195–217.

Hussong, D, Colwell, RR, O'Brien, M, Weiss, E, Pearson, AD, Weiner, RM and Burge, WD (1987) Viable *Legionella pneumophila* not culturable by culture on agar medium. *Bio/Technology* **5**:947–950.

Kaprelyants, AS and Kell, DB (1996) Do bacteria need to communicate with each other for growth. *Trends in Microbiology* **4**:237–242.

Kotiaho, T, Lauritsen, F, Degn, H and Paakkanen, H (1995) Membrane inlet ion mobility spectrometry for on-line measurement of ethanol in beer and in yeast fermentation. *Analytical Chimia Acta* **309**:317–325.

Kurokawa, M, Nukina, M, Nakanishi, H, Tomita, S, Tamura, T and Shimoyama, T (1999) Resuscitation from the viable but nonculturable state of *Helicobacter pylori*. *Kansenshogaku Zasshi* **73**:15–19.

Little, B, Wagner, P and Mansfield, F (1991) Microbiologically influenced corrosion of metal and alloys. *International Material Reviews* **36**:253–272.

Lleo, MdM, Tafi, MC and Canepari, P (1998) Nonculturable *Enterococcus faecalis* cells are metabolically active and capable of resuming active growth. *Systematics in Applied Microbiology* **21**:333–339.

Melo, L and Pinheiro, M (1992) Biofouling in heat exchangers. In: Biofilms—Science and Technology (Eds. Melo, L, M, f, Bott, T and Capdeville, B), Kluwer Academic Publishers, Dordrecht, pp. 499–509.

Nealson, KH, Platt, T and Hastings, JW (1970) Cellular control of the synthesis and activity of the bacterial luminescent system. *Journal of Bacteriology* **104**:313–322.

Nichols, PD, Henson, JM, Guckert, JB, Nivens, DE and White, DC (1985) Fourier transform-infrared spectroscopic methods for microbial ecology:analysis of bacteria, bacterial polymer mixtures and biofilms. *Journal of Microbiological Methods* **4**:79–94.

Nivens, DE, Chambers, JQ, Anderson, TR, Tunlid, A, Smit, J and White, DC (1993a) Monitoring microbial adhesion and biofilm formation by attenuated total reflection/fourier transform infrared spectroscopy. *Journal of Microbiological Methods* **17**:199–213.

Nivens, DE, Chambers, JQ, Anderson, TR and White, DC (1993b) Long-term on-line monitoring of microbial biofilms using a quartz crystal microbalance. *Analytical Chemistry* **65**:65–69.

Nivens, DE, Schmitt, J, Sniatecki, J, Anderson, TR, Chambers, JQ and White, DC (1993c) Multi-channel ATR/FT-IR spectrometer for on-line examination of microbial biofilms. *Applied Spectroscopy* **47**:669–671.

Paakkanen, H and Huttunen, J (1994) Gas detector based ion mobility. *Finnish Air Pollution Prevention News* **20**:20–23.

Pearson, JP, Gray, KM, Passador, L, Tucker, KD, Eberhard, A, Iglewski, BH and Greenberg, EP (1994) Structure of the autoinducer required for expression of *Pseudomonas aeruginosa* virulence genes. *Proceedings of the National Academy of Science USA* **91**:197–201.

Rollins, DM and Colwell, RR (1986) Viable but nonculturable stage of *Campylobacter jejuni* and its role in survival in the natural aquatic environment. *Applied and Environmental Microbiology* **52**:531–538.

Roszak, DB and Colwell, RR (1987) Survival strategies of bacteria in the natural environment. *Microbiological Reviews* **51**:365–379.

Roszak, DB, Grimes, DJ and Colwell, RR (1984) Viable but nonrecoverable stage of *Salmonella enteritidis* in aquatic systems. *Canadian Journal of Microbiology* **30**:334–338.

Steinert, M, Emody, L, Amann, R and Hacker, J (1997) Resuscitation of viable but nonculturable *Legionella pneumophila* Philadelphia JR32 by *Acanthamoeba castellanii*. *Applied and Environmental Microbiology* **63**:2047–2053.

Stoodley, P, deBeer, D and Lappin-Scott, HM (1997) Influence of electric fields and pH on biofilm structure as related to the bioelectric effect. *Antimicrobial Agents and Chemotherapy* **41**:1876–1879.

Tamachkiarowa, A and Flemming, H-C (1996) Glass fibre sensor for the monitoring of biofouling. *Dechema Monography* **133**:31–36.

Williams, P and Stewart, G (1994) Cell density dependent control of gene expression in bacteria—implications for biofilm development and control. In: Bacterial biofilms and their control in medicine and industry (Eds. Wimpenny, J, Nichols, W, Stickler, D and Lappin-Scott, H), Bioline, Cardiff, pp. 9–12.

Index